Biology

SCIENTIFIC PROCESS AND SOCIAL ISSUES

Garland E. Allen
Washington University

Jeffrey J.W. Baker
Formerly of Wesleyan University

FITZGERALD SCIENCE PRESS, INC.
Bethesda, Maryland

Fitzgerald Science Press, Inc.
Editorial Offices
4814 Auburn Ave.
Bethesda, Maryland 20814
(301) 654-6884
www.fitzscipress.com

Fitzgerald Science Press, Inc.
Orders and Fulfillment:
P.O. Box 605
Herndon, VA 20172
(800) 869-4409
Fax: (7003) 661-1501

Between the time Website information is gather and published, some sites may have closed or changed. Also, the transcription of URLs may result in typographical errors. Fitzgerald Science Press would appreciate notification where these occur so that they may be corrected in subsequent editions.

Library of Congress Cataloging-in-Publication Data

Allen, Garland E.
  Biology : scientific process and social issues / Garland E. Allen, Jeffrey J.W. Baker.
    p. cm.
Includes bibliographical references
  ISBN 1-891786-09-1
  1. Biology--Methodology. 2. Biology--Social aspects. I. Baker, Jeffrey J. W. II. Title.
QH324 .A48 2001
570--dc21

                                        2001023963
Editor: Nancy Knight
Art: Network Graphics
Production Management: Chestnut Hill Enterprises, Inc.
Composition: The Format Group LLC
Interior Design: Susan Brown Schmidler
Cover Design: Rock Creek Publishing
Printer: United Book Press

01 02 03 04—10 9 8 7 6 5 4 3 2 1

To Larry, Tania, and Carin, who have taught me a lot about process,

y

a Barbarita de Vieques, con mucho respeto, amor, y pesar.

Contents

Preface   xi

1 Biology as a Process of Inquiry   1
The Growth of Biological Thought: The Nineteenth Century   2
The Twentieth Century Revolution in Biology   8
The Spirit of Inquiry: Some Unsolved Problems in Biology   14
  Butterfly Migration   17
  Hens' Teeth and Evolution   22
  The Origin of Life   24
Conclusion   29
Further Reading   29
Suggested Websites   31

2 The Nature and Logic of Science   33
Science as an Intellectual Discipline   34
  Defining Science   34
  Characteristics of Science   34
  The Relationship Between the Science, Social Sciences, and Humanities   36
  Science, Common Sense, and Intuition   37
The Logic of Science   37
  Observation and Fact   37
  From Fact to Conceptualization   39

Types of Conceptualizations   40

Observation, Fact, and Conceptualization: A Case Study   41

Creativity in Science   44

The Logic of Science: Induction and Deduction   45

Induction and Deduction   45

Logic, Predictions, and the Testing of Hypotheses   47

The Concept of Proof in Science   48

The "Dissection" of an Experiment   49

The Logic of Science: Hypotheses as Explanations   50

Teleological vs Causal Hypotheses   50

Types of Causal Explanations   55

Cause and Effect   57

Bias in Science   57

Conscious Bias   58

Unconscious Bias   59

The Concept of Paradigms   66

Darwin and the Theory of Evolution by Natural Selection   67

*A Paradigm Shift in Molecular Biology: From the "Central Dogma"
to Reverse Transcription   69*

Characteristics of Paradigms   72

Modern Science, Materialism and Idealism   74

Mechanism and Vitalism   74

Conclusion: The Strengths and Limitations of Science   80

Exercises for Critical Thinking   81

Further Reading   84

Suggested Websites   85

**3   *The Nature and Logic of Science: Testing
Hypotheses   87***

Testing Hypotheses: Some General Principles   87

Testing Hypotheses by Observation   87

Testing Hypotheses by Experiments   88

The Importance of Uniformity and Sample Size   90

A Case Study in Hypothesis Formulation and Testing:
How Is Cholera Transmitted?   92

Disease: The Early Years   92

Snow's Observations 93

An Alternative Hypothesis 95

The Case of the Broad Street Pump 97

Objections to Snow's Water-Borne Hypothesis 97

The Critical Test 99

A Wider Appllicability and the Role of Chance 100

The Mechanism of Cholera Action: A Modern Perspective 101

A Modern Epidemic: The AIDS Crisis, 1981–? 101

Background: Origin of the AIDS Crisis 101

The Nature of HIV Infection 106

Treatment and/or Cure for AIDS 110

Conclusion 111

Exercises for Critical Thinking 112

Further Reading 113

Suggested Websites 113

4 *Doing Biology: Three Case Studies* 115

Formulating and Testing Hypotheses in the Laboratory: The Discovery of Nerve Growth Factor 116

Research in the Field: Homing in Salmon 122

The Organism 123

Two Possible Hypotheses 124

Tabulating the Results 126

Additional Questions 128

Testing the Pheromone Hypothesis Further 132

Dissecting the Experiments 134

Statistical Significance 135

Research in Historical Biology: Mass Extinction and the End of the Dinosaurs—The Nemesis Affair 138

Testing the Impact Hypothesis 139

Extinction and the Paleontological Record 140

Periodic Mass Extinctions 142

The Nemesis Hypothesis 144

Conclusion 146

Exercises for Critical Thinking 146

Further Reading 147

Suggested Websites 147

5 The Social Context of Science: The Interaction of Science and Society   149

Science and Technology: The Public Confusion   150

The Social Construction of Science   151

The Origin of The Origin of Species: Darwin and the Political Economists   152

Homeostasis: Walter Bradford Cannon and the New Deal   155

The Social Responsibility of Science   158

Genetics and Eugenics in the Early Twentieth Century   159

Herbicides in Southeast Asia   162

Ethical Concerns in Science Today   163

Ethical and Social Issues in the Use of Human Subjects for Research   164

The Tuskegee Study   164

Science and Pseudoscience   168

"Scientific" Creationism   169

Some Historical Background   169

Creationism: What Is It?   170

"Scientific" Creationism vs Evolution   171

Creationism as a Science   175

Science and Religion: An Underlying Difference   181

Conclusion   182

Exercises for Critical Thinking   183

Further Reading   184

Suggested Websites   185

Appendix 1 The Analysis and Interpretation of Data   187

Collecting Data and the Problem of Sampling Error   188

Seeking Relationshops: Collecting and Organizing Data   188

Quantitative and Qualitative Data   188

Measurement and Precision in Data Gathering   189

Variations in Measurement by Different Observers   189

Seeking Relationships: The Presentation of Data   190

Displaying Data   190

Scales and Scalar Transformation   194

Interpolation and Extrapolation   194

The Analysis and Interpretation of Data   196
  Correlation   196
  Rate and Change of Rate   201
Analysis of Distributions: Central Tendency and Dispersion   201
  Levels of Significance   204
Further Reading   207

Appendix 2 Exercises for Critical Thinking:
Possible Answers   208

Index   215

About the Authors   223

Preface

This book grew out of our many years of teaching introductory level biology and related science courses at several different colleges and universities, primarily at Wesleyan University, in Middletown, Connecticut (Baker), and Washington University, in St. Louis, Missouri (Allen). We envision this book as a core text or supplementary text in courses for both biology majors and nonmajors. Its format is designed specifically to address the nature of science in general and of the biological sciences specifically. We hope that this brief presentation of the process of doing biology as a science will complement a more traditional, fact-driven presentation, whether in the form of any of the excellent comprehensive biology texts now available or a compendium of resources from the World Wide Web.

We remember from early years of college teaching the distinct feeling that, more often than not, some of our most creative students received only mediocre grades. Unmotivated by the presentation of a dazzling array of detailed factual information, they would often ask unexpectedly thoughtful and highly reflective questions. Conversely, students receiving As, although able to regurgitate biology "facts" (for example, details of the Krebs citric acid cycle), were often totally incapable of mapping out a strategy for even the simplest of research projects. A variation on this theme was the comment of one faculty colleague that he found himself becoming increasingly uncomfortable with the recognition that some of his students could complete a basic introductory biology course satisfactorily yet remain convinced of the intellectual validity of Creationism!

These and other such experiences convinced us that students may do well in a science course but fail completely to comprehend the nature of science itself, the underlying reasons for its immense power as an intellectual process, and the relationship between the natural sciences, social sciences, and humanities. Worse, students seem to drift from one extreme to another when considering science, either buying into popular press accounts of science as the ultimate source of all truth or viewing it with suspicion, fear, and even hostility. Unfortunately, this latter view is often found even among the ranks of college faculty in the social sciences and the humanities, where science is sometimes referred to as "just another system of myths," "the predominant myth of the twentieth century," or other words to that effect. Such a view does a profound disservice to the meaning of both "science" and "myth." If, indeed, modern science is to be viewed as a myth, then clearly it is by far the most powerful one yet devised by the human imagination!

It was in an attempt to deal with this confusion that, over the years, we began to incorporate into our teaching material designed to confront this and other such problems. In one of our earlier books (*The Study of Biology*, 4th. ed. Reading, MA: Addison Wesley Publishing Company, Inc.; 1984), we put forward what we think is the essence of a proper approach to science: ". . . it is the process, not merely the content or "facts," of biology that should form the basic thrust of introductory courses. By the process of science we mean how we know what we know; how experiments are designed, data analyzed, and conclusions drawn designed to stress to students that today's scientific "facts" may very well be tomorrow's errors. Therefore, when Fitzgerald Science Press expressed interest in the underlying philosophy of our work and suggested publishing it as a brief text suitable for a wide range of life science courses, we felt that a genuine meeting of the minds had occurred.

Overview of the Contents

Chapter 1 explores the historical development of the life sciences and looks ahead to what the field might be like as we start the new millenium. Current examples of research in the field are used to stress that biology, like all the natural sciences, has many unsolved problems that researchers today are still trying to understand. The field is no less dynamic now than in the past.

Chapter 2 begins with an in-depth look at biology as a science in the context of hypothesis formulation and underlying inductive and deductive logic. Here we stress the differing types of hypotheses, the roles that conscious and unconscious bias may play in both formulating questions and evaluating answers, and the strengths and limitations of science as an intellectual discipline.

Chapter 3 examines the ways in which scientific hypotheses are tested and how their logic may be analyzed in terms of their inductive and deductive framework. We have often elected here and elsewhere in this book to use older studies from the seventeenth, eighteenth, and nineteenth centuries. So doing enables us to avoid having to provide a great deal of subject matter background and, because the student reader more than likely already knows the "answer," he or she is forced to concentrate instead on the underlying intellectual structure of the science involved. We then apply insights from these older case studies to contemporary issues: for example, conflicting hypotheses concerning the cause of the AIDS epidemic now sweeping many developed and developing nations. For instructors interested in introducing the concept of statistical analysis in hypothesis evaluation, we provide an appendix dealing with some of the essential concepts, techniques, and means used to establish correlation, standard deviation, and levels of confidence.

Chapter 4 compares three case studies, the first an example of research carried out in the laboratory and the second an example of research in the field, with all the difficulties in terms of controlling variables that such studies entail. The third case study deals with problems inherent in evolutionary studies, in which events have occurred in the past and so cannot be observed directly.

Finally, Chapter 5 addresses the interrelationship of science and the society within which it develops and the social responsibility of scientists to the society that funds their research. Using Creationism as our example, we end with issues raised by the still all-too-powerful attraction of pseudoscience and its misuse to further specific political agendas. Instructors doubtless may wish to bring in other examples of their own.

Appendix 1 provides a succinct review of some basic concepts in statistics, data analysis, and experimental design relevant to the many cases discussed in the text.

The last four chapters conclude with exercises for the reader that call for critical thinking about the issues presented. Possible answers, with stress on the word "possible," are suggested in Appendix 2.

Acknowledgments

We wish to thank the manuscript reviewers for their many valuable comments and suggestions: Ann Burgess, University of Wisconsin; Neil Campbell, University of California, Riverside; Geoffrey Cooper, Boston University; Robert Huskey, University of Virginia; Tony Lawson, Arizona State University; Karen Ocurr, University of Michigan; Gail Patt, Boston

University; Steve Rissing, Ohio State University; Ralph Sorensen, Gettysburg College; Ruth Welti, Kansas State University; and Larry Wiseman, College of William and Mary. Their many helpful comments and suggestions aided us greatly in improving both clarity of expression and in sorting out the essential from the trivial. For the most part, we followed their suggestions but, because we did not always do so, we and we alone are responsible for this final version. A special note of thanks also to Ralph Sorensen for preparing the Web links that appear at the end of chapters.

We are also indebted to our former collaborator, Scott F. Gilbert of Swarthmore College, for his early enthusiasm in the project that eventually led to this book. The "hen's teeth" example was first suggested by him, as well as the idea for a chapter on "unsolved problems" in modern biology.

It is also only appropriate that we express here our gratitude for the support and able assistance provided us by Irma Morose, Teresa L. Tate (née Lowe), and Cindy Marks. Special thanks also to our editor, Nancy Knight, for her incredibly high level of professional competence and for sharing with us via email an absolutely marvelous sense of humor! Patrick Fitzgerald's keen interest in this book and his obvious enthusiasm for it was both an inspiration and a spur, as he grumpily pushed us hard to meet his deadlines. All too often such valuable input goes unrecognized, despite the fact that little or nothing could be accomplished without it.

We touch in this book upon the recent triumph of human genome sequencing that was announced jointly by two major groups of scientists as this text went to press. The resulting recognition of bacterial DNA sequences embedded within the human genome represents a stunning victory both for human ingenuity and biology's underlying paradigm, that of evolution, with its beautifully simple and inspiring concept of the unity and interrelatedness of all living organisms. Beyond this thrilling intellectual achievement, there is the extraordinary future potential of the field of biology and the central role it must play in addressing such issues as human health and disease and the impact of global environmental changes. It is truly a great time in human history to be a biologist . . . or, perhaps, to consider becoming one.

Garland E. Allen, *St. Louis, Missouri*
Jeffrey J. W. Baker, *Ivy, Virginia*
February, 2001

Biology as a Process of Inquiry

I magine yourself on a moonlit night in mid-August at the Marine Biological Laboratory in Woods Hole, Massachusetts. You are standing on a pier looking into the shallow waters below. The reflected moonlight rippling through the water gives the appearance that the whole ocean is alive. As your eyes adjust to the darkness you realize that the water is teeming with life. You are observing the mating ritual, triggered by the full moon, of thousands of writhing, swirling polychete worms. A timeless event unfolds before you, drawing you back to primeval oceans where ancient progenitors of these polychetes repeated a similar ritual, long before any human eyes were there to observe. Here is an expression of life at its most fundamental, dynamic, and breathtaking. You are witnessing a great saga: past, present, and future all rolled into one spectacular moment.

Many who have witnessed this or similar events around the world have found in the activities of living organisms an inspiration to study, investigate, and learn more about this wonderful phenomenon we call "life." Biology, the scientific study of life, builds on this inspiration to ask questions: Why, for example, does the polychete mating process occur only in late summer? How is it timed, and what is it about the phases of the moon that trigger the mating behavior? Without this awe and wonder, there can be no sustained investigation. Without sustained inquiry, we look on as outsiders and do not gain access to life's mysteries. *Sustained inquiry is what biology as a science is all about.*

This book is about how biologists approach the study of life. How do we ask the right questions, questions to which it is possible to obtain answers? What distinguishes a useful from a not-so-useful question or one that can be tested from one that cannot? What are the respective roles of observations, facts, hypotheses, and theories in science? How are experiments designed, carried out, and the resulting data analyzed? These and other issues implicit in asking what science is and how we know what we know make up a central theme of this book.

The Growth of Biological Thought: The Nineteenth Century

If the nineteenth century witnessed the rise to preeminence of the physical sciences, the twentieth century witnessed a similar rise of the biological sciences. As we enter the new millennium, biology represents one of the most rapidly developing and exciting areas of scientific investigation the world has ever known.

Throughout most of its earlier history, biology was a largely descriptive science, concerned with topics such as comparative anatomy and embryology, descriptions of the adaptations of organisms to their environments, and the classification of species. Most early biologists had strong backgrounds in natural history and field work. They studied their collected specimens in museums and laboratories and generally did not conduct experiments of the sort carried out in chemistry and physics laboratories. It was for these reasons that biology was considered to be a distinctly second-class citizen by those members of the academic community working in natural sciences such as physics, chemistry, and geology.

Two areas of biological thought underwent major developments in the middle and latter part of the nineteenth century: physiology and the study of evolution. From the early 1850s on, both German and French physiologists pioneered the application of physical and chemical methods to the study of the ways in which organisms function. For example, Hermann von Helmholtz (1821–1894), in Germany, found ways to measure the speed of nerve impulses and to demonstrate their electrochemical nature. Claude Bernard (1813–1878), in France, studied the chemistry of the liver and demonstrated its ability to manufacture animal starch from simple sugars, the first time such a synthesis had been demonstrated in animals (it was well known in plants by this time). However, the most momentous scientific development in nineteenth-century biology was undoubtedly the publication in 1859 of *On the Origin of Species by Means of Natural Selection*, by Charles Darwin (1809–1882). Darwin not only promoted the idea, long well known but not well accepted, of evolution (or "transmutation of species," as it was called) but also put forward the theory of natural selection, a mechanism by which evolution might occur. He noted that in nature many more offspring are born than ever reach maturity, the latter a result of competition for limited resources. Any genetically-based variation an organism possesses could give it a slight edge in this competitive struggle. That "edge" would be favored by natural selection if it in any way allowed the organism to leave more offspring that could reach reproductive maturity, than other organisms lacking the variation. Thus favorable variations would spread throughout the population over successive generations, unfavorable variations would be eliminated, and the character of the population changed over time: that is, evolution would have occurred. Although Darwin's theory originally was controversial among biologists and nonbiologists alike, it has survived to become the most comprehensive organizing concept of modern biology.

Following Darwin's lead, biologists devoted themselves to working out descriptive aspects of evolutionary theory, and, as noted earlier, it was this aspect of biology that came to dominate much of the work in the later nineteenth century. An example of this descriptive biology was a field known as morphology, literally the "study of form." In the 1870s and 1880s, morphology was devoted primarily to tracing evolutionary relationships based on evidence from comparative anatomy, embryology, physiology, and the fossil record. The end product of such morphological work was the production of phylogenetic "trees" (Fig. 1.1). For example, to determine whether a group known as annelids (earthworms and polychetes such as those described at the opening of this chapter) and arthropods (insects, crayfish, spiders) might share a common ancestor, morphologists would compare the anatomy of the adult and embryonic forms. It was noted that both groups were segmented, possessed similar nervous and circulatory systems, and developed from fertilized eggs in the same pattern of cell division and growth. This morphological similarity was strengthened by finding an animal, *Peripatus* (Fig. 1.2), still living and sharing characteristics of both groups. Like the annelids, *Peripatus* is composed of repeated and nearly identical segments but has jointed appendages on each segment, as do arthropods. *Peripatus* thus was believed to be the descendant of a form intermediate between the annelids

A   PEDIGREE OF MAN.

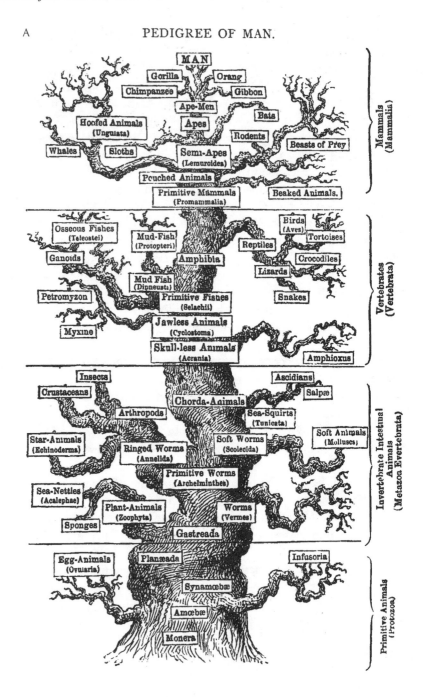

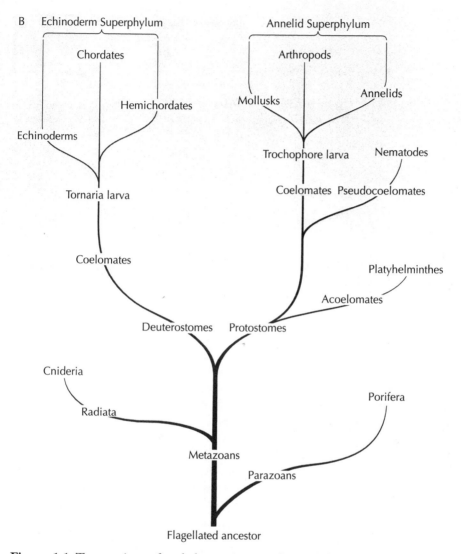

B   Echinoderm Superphylum   Annelid Superphylum

Chordates   Arthropods

Hemichordates   Mollusks   Annelids

Echinoderms

Tornaria larva   Trochophore larva   Nematodes

Coelomates Pseudocoelomates

Coelomates

Platyhelminthes

Acoelomates

Deuterostomes   Protostomes

Cnideria

Porifera

Radiata

Metazoans

Parazoans

Flagellated ancestor

Figure 1.1 Two versions of a phylogenetic tree, showing the evolutionary relationships thought to exist between the major animal groups (phyla). (A) A nineteenth-century version, drawn for morphologist Ernst Haeckel's *The Evolution of Man* (1879). The tree shows not only the fascination of early morphologists with reconstructing the evolutionary relationships among organisms but also their straight-line, progress-oriented concept of evolution. The vertical tree trunk is shown as growing toward the ultimate product of evolution, human beings. Many of the relationships depicted here were based on evidence derived from embryological studies. (B) A more recent evolutionary tree. Note that it is more bush-like, with no single progression leading toward any one group.

Figure 1.2 *Peripatus*, a member of the phylum Onycophora, is thought to be a modern-day descendant of a form ancestral to both the arthropods and annelids. The discovery of living or fossil forms intermediate between two quite divergent modern groups provides strong evidence for the Darwinian concept of evolution by natural selection. © Cabisco/Visuals Unlimited.

and arthropods. To morphologists it represented something like what the common ancestor might have resembled.

A limitation of morphological work was that there were no methods available at the time with which to test many of the ideas concerning evolutionary relationships. Thus the various scenarios remained highly speculative. (Such a limitation, of course, does not invalidate the attempt to determine evolutionary relationships by comparative methods. Indeed, such efforts have gained a considerable following today, using a variety of molecular data unavailable in the past.) A break from the morphological tradition occurred in the 1880s and 1890s, when German and French biologists became interested in studying the embryo for its own sake rather than only to uncover clues about phylogeny. These biologists also were highly attuned to testing their ideas experimentally. For example, German embryologist Hans Spemann (1869–1941) became interested in how the fertilized egg, after a number of cell divisions (cleavages), produced a variety of different cell types (muscle, nerve, bone, etc.). A common view at the time, known as the "mosaic theory," maintained that different parts of the egg contained different determiners or factors, which, as the egg divided, became parceled out to different daughter cells. A cell eventually would develop into a specific type according to the kind of determiner or factor it ended up containing.

Spemann worked with amphibian eggs (mostly salamander), because they were plentiful, large enough to be manipulated experimentally, and could be kept alive in the laboratory with relative ease. Most of the fertilized salamander egg looks like a uniform mass of cytoplasm, except for one noticeably different region, the gray crescent. Spemann suspected that this area of the egg might have something to do with directing the first phases of embryonic development. The salamander egg normally divides along a plane, ensuring that each daughter cell gets part of the gray crescent. In a series of experiments using fine baby hair (from his infant son's head), Spemann carefully tied dividing salamander eggs in their normal plane of division, so that the cells on each side of the constriction contained gray crescent material (Fig. 1.3A). Each part of the cells divided and gave rise to more or less complete (though dwarf) embryos. Spemann reasoned that if material from the gray crescent was necessary for complete development, then tying the dividing egg in a plane perpendicular to the plane of cell division, so that one side of the constriction received all the crescent whereas the other side received none, should result in only that part of the cell containing gray crescent material developing into a complete embryo.

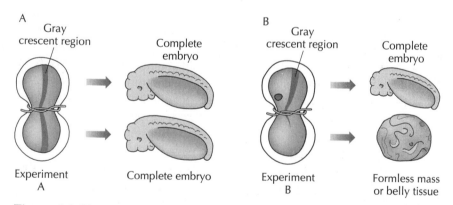

A
Gray
crescent region
Complete
embryo

Experiment
A
Complete embryo

B
Gray
crescent region
Complete
embryo

Experiment
B
Formless mass
or belly tissue

Figure 1.3 Hans Spemann's experiments showing the importance of the gray crescent in the ability of the first two cells of an embryo to develop normally. (A) When Spemann experimentally constricted a dividing salamander egg so that gray crescent material was present in both cells, each cell would develop normally into a complete, though dwarf, embryo. (B) If the constriction was made in the same plane as the gray crescent, so that one side of the constriction received all the gray crescent material, only that cell produced a normal embryo. The other cell developed into a formless mass of cells. These experiments showed that specific determinants associated with the gray crescent were necessary for full development.

When he performed this experiment (Fig. 1.3B), his prediction was borne out. The part of the cell lacking gray crescent material produced only a formless mass of cells, whereas the region containing the gray crescent formed a complete embryo. To Spemann, these results were conclusive evidence that the way in which material—in this case, gray crescent material—was parceled out to daughter cells determined the overall course of development. This demonstrated experimentally that, in ways unknown at the time, the organization of the fertilized egg had an important effect on the future course of embryonic development.

As in the physical sciences, experiments such as Spemann's (for this and other work he was awarded a Nobel prize in 1935) allowed biologists to obtain precise answers. It was through such experiments that biology began to acquire an image of being more rigorous and to come into its own among the other experimental sciences.

The Twentieth-Century Revolution in Biology

The rise of biology to its new scientific status coincided with the beginning of the twentieth century. In the year 1900, the original work on heredity by Augustinian monk Gregor Mendel (1822–1884) was rediscovered, and, as a result, the new science of genetics was born. By 1950, genetics had become central to virtually all areas of modern biology. Questions about the hereditary units called genes and how they functioned led first to determination of the relationship between genes and the synthesis of proteins. Eventually these questions would lead to determination of the double-helix structure of the molecule of heredity itself, deoxyribonucleic acid (DNA), by James D. Watson and Francis Crick. Genes now were recognized as consisting of DNA. From this work arose the entire field of molecular biology, the study of the atom-by-atom structure of biologically important molecules. The subsequent development of genetic engineering, including the cloning of agriculturally important animals and plants, has opened up new potentials for food production and management. Many genetic diseases, once believed to be incurable, are now open to the prospect of drug or gene therapy.

Perhaps most significant, genetics provided the mechanism of heredity that Darwin's theory of evolution had lacked, leading to new advances in understanding how species evolve over time. One of the weak points in the theory of natural selection as originally proposed by Darwin was that he had no clear idea how heredity worked. Like most of his contemporaries, he thought that characteristics acquired by an organism during its lifetime, especially changes induced by diet, temperature, and changed habits, could be inherited by its offspring. He also thought that two different forms of the same trait (for example, red and white flower color) would produce a blend in the offspring, thus diluting the trait in successive generations. In the late 1920s and early 1930s, various mathematically oriented biologists began applying Mendelian genetics to the evolutionary process, creating the field of population genetics. Since then, evolutionary biologists have worked out the dynamics of the ways in which natural selection works in a large variety of animal and plant populations. With its marriage to Mendelian genetics, the Darwinian theory of evolution by natural selection became thoroughly established as *the* central concept underlying all of modern biology and has become one of the most active areas of modern biological research.

Two aspects of evolutionary biology commanded the attention of biologists throughout much of the last half of the twentieth century. One is molecular evolution, the attempt to understand how evolution works at the molecular level. DNA guides the assembly of proteins, with each DNA segment determining a very specific protein structure. As a result, changes in DNA (mutations) are often reflected in variations in the structure and/or function of protein molecules. Because protein structural differences can be determined with great precision, it is possible to test more rigorously hypotheses about evolutionary relatedness, by comparing the same protein in different species. The same can be done with DNA directly. Such studies have shown that proteins exist in "families" (for example, cytochromes, proteins involved in cellular respiration). Like organisms themselves, the members of these families may be thought of as having diverged from a common ancestral form. Proteins from very different families are now known to be composed in many instances of similar motifs or patterns that are "highly conserved," that is, used repeatedly throughout evolutionary history.

Yet another aspect of evolutionary biology that made great strides in the twentieth century was the study of human evolution. Excavations at Olduvai Gorge in East Africa in the 1950s and 1960s by Louis Leakey (1903–1972) and Mary Leakey and in the Afar region of Ethiopia by Donald Johanson in the 1970s brought to light fossil hominids (human-like forms) much older than any discovered previously. These finds suggested, among other things, that our ancestors walked upright long before our brain size expanded (or, as has been said, our evolution was feet first rather than head first!). New finds now are made almost every year. From computer-based studies of modern populations, physical anthropologists have concluded that it is likely that our own genus and species (*Homo sapiens*) originated in Africa and later migrated to all areas of Asia and Europe. Our closest relative in time, the Neanderthals (*Homo neanderthalensis*) now are believed to have roamed Europe as recently as 28,000 years ago. This much more recent estimate of their existence indicates that Neanderthals must have overlapped with our own species. Such discoveries have raised intriguing questions about how similar the Neanderthals were to us and what caused them to become extinct.

Toward the end of the twentieth century, biologists embarked on their most ambitious project yet: the Human Genome Project (HGP), sometimes also referred to as the Human Genome Initiative. A large-scale international effort, the HGP's primary goal has been to map and sequence all of the base pairs in the DNA of the human genome, a term referring to the complete set of genes contained in an organism. The purpose of this effort was to identify first the structure of all the genes and then determine their functions. Many different laboratories in the United States and Europe had been working since the early 1990s to complete the sequences. The technology developed for this work increased dramatically the rate at which sequencing could be carried out. One result was the isolation and molecular characterization of genes for several important human diseases, such as cystic fibrosis (a progressive build-up of mucus on the tissue lining the lungs and intestines leading, in many cases, to premature death) and Huntington's chorea (a gradual deterioration of the nervous and muscular systems). Genes affecting other rare diseases already have been mapped to specific regions of chromosomes, and many of these have been sequenced.

To help in understanding the functions of these sequences, the genomes of several other experimental organisms have been or are being analyzed: yeast (completed), the small roundworm *Caenorhabditis elegans* (completed), the thale cress *Arabidopsis thaliana* (completed), the mouse, and the fruit fly *Drosophila* (completed). What already has emerged from this comparative effort is the recognition that many of the same DNA sequences appear in a wide range of organisms. For example, the homeotic genes constitute one frequently encountered group of genes in the animal kingdom. These genes appear to be responsible for such fundamental processes in early embryonic development as establishing the anterior-posterior axis of the organism and the formation of the body segments found in insects and the vertebral column of vertebrates. Just as an architect uses a few basic modular blueprints that may be combined in different ways to produce quite different buildings, evolution has preserved batteries of genes that, in conjunction with other DNA sequences, build very different organisms using many of the same starting instructions.

Yet another area of biology that showed enormous growth in the twentieth century was immunobiology, a field investigating how the immune system functions at the systemic, cellular, and molecular levels to protect the body from invasion by foreign agents such as bacteria and viruses. Spurred on, in part, by new concepts in developmental biology in the 1960s, study of the immune process has expanded enormously in the past decade in response to the acquired immune deficiency syndrome (AIDS) epidemic rampant throughout the world. AIDS, a disease marked by a decline in the ability of immune system cells (T cells) to respond to infectious agents, is caused by a virus that has ribonucleic acid (RNA), instead of DNA, as its genetic material. People with AIDS lose the ability to fight off not only the AIDS virus itself but many other infectious agents as well. Attempts to understand the mechanism of infection and replication by the AIDS virus and the response to it by the antibody-producing cells of the immune system have led to an explosion in research in the fields of virology and immunology. Drugs that interfere with the replication of the AIDS virus have revealed many details about the process of viral infection and host cell response. Although a cure is still in the future, considerable progress has been made in devising combinations of drugs that slow and even appear to stop the replication and spread of the AIDS virus within the host's body.

Other areas of biology that flourished and expanded our horizons greatly during the twentieth century were neurobiology, cell biology, and ecology. Neurobiology is a broad umbrella field that covers everything from the study of how individual neurons function to such issues as the nature of brain function, memory, and consciousness. The twentieth century opened with a debate about whether the nervous system was composed of individual cells (neurons) or was a continuous, fibrous network (reticulum), a debate resolved in favor of the neuron. Major advances were made between the 1930s and 1960s in devising techniques for studying single neurons and recording electrochemical changes inside and outside the cell. The marine squid (*Loligo peali*) figured prominently in this work because of its widespread distribution and the very large neuron running along its dorsal (upper) surface. So large is this neuron that it is possible to insert one microelectrode inside the cell and keep another on the outside, enabling the recording of voltage changes across the membrane as the neuron conducts an impulse.

Along the same lines, much also has been learned about the process of transmission between neurons. Substances called neurotransmitters, which are released from one cell and stimulate (or inhibit) the other neurons with which it forms a junction, have been shown to play very important roles in a wide variety of behaviors, including the regulation of human emotions.

The relationship between neuronal activity and behavior has been studied in exquisite detail in a variety of organisms, including the small roundworm and the sea slug (*Aplysia*). Both of these organisms have simple nervous systems in which virtually every neuron is known and has been mapped. Genetic or experimentally induced defects in specific neurons in the network have been correlated with specific behavioral changes and provide a glimpse into how simple nervous systems are organized and how they function in making specific responses. Top among the findings during the last two decades of the century was the role of a special class of neurotransmitters, the biogenic amines, including dopamine, epinephrine, norephinephrine, and serotonin. These neurotransmitters have been found to play roles in increasing or inhibiting certain inputs to the brain and central nervous system. Serotonin, for example, is produced in highest quantities when the organism is awake and alert and in lowest quantities during sleep. In chimpanzees, serotonin levels change in response to changes in the animal's social position: Increase in social position brings about a decrease in overall serotonin levels and vice versa.

The molecular bases of our conscious and altered states of mind, as well as of memory and thought, are increasingly yielding to biochemical analysis. This does not mean we are simply biochemical robots, because ever-changing inputs from the environment are constantly altering our physiological state. Although our biology certainly determines our behavior, our behavior also determines our biology. The relationship is not a simple one, and overstated claims that complex behavior and personality traits are the products of one or two neurotransmitters or genes should be regarded with suspicion.

Our understanding of the structure and function of cells, the basic units of life, underwent a genuine revolution during the past century. In 1896, Edmund Beecher Wilson (1856–1939) at Columbia University published the first edition of his book, *The Cell in Development and Inheritance*, which was to become a classic of cell biology throughout the first half of the twentieth century. Wilson's summary of cell structure and function of both adult and embryonic cells was derived mostly from observational studies with the light microscope. At the time, cells were still thought of as membrane-bound units with several specific internal structures, such as the nucleus, chromosomes, and vacuoles, suspended in an amorphous medium known as "protoplasm". With the advent of the electron microscope in the

1930s and expanding use after World War II, protoplasm turned out to possess a complex structure of internal membranes; particle-like structures such as ribosomes (the sites of protein synthesis); organelles ("little organs"), such as mitochondria and chloroplasts, each with its own complex structure; and a variety of structural components known collectively as the cytoskeleton. By the end of the century, study of the structural and functional properties of cells using still more sophisticated imaging, such as the confocal microscope, added further detail to our understanding of the molecular biology of the cell.

At the macroscopic (large) rather than microscopic (small) level of ecological processes, biologists have made some remarkable advances in the twentieth century. The field of ecology was created in the second decade of the century, and the study of the world's major ecosystems (communities of organisms that interact in the same geographic space) has proceeded with great rapidity ever since. Yet, as we have learned more about the intricate interconnections that occur in most ecosystems, we have also learned how fragile they are. Human activities have had a major impact on ecosystems as diverse as the Atlantic Ocean and the tropical rain forests of South America. Writing in 1999, Peter Raven, Director of the Missouri Botanical Garden, Professor of Biology at Washington University (St. Louis, MO), and an energetic spokesperson for the preservation of ecosystems and their diversity, emphasized this point when he stated that:

> The most serious and rapidly accelerating of all global environmental problems is the loss of biodiversity. Over the past 300 years many species of organisms, including mammals, birds, and plants have been lost. In addition, habitat is vanishing rapidly, especially in the tropics. Scientists...have calculated that as much as 20% of the world's biodiversity may be lost during the next 30 years. No more than 15% of the world's eukaryotic organisms have been described, and a much smaller proportion of tropical organisms have been named. Thus, we may never even know of the existence of many of the organisms we are driving to extinction.... This is a tragic loss for several reasons. First, many feel that on moral, ethical, and aesthetic grounds, we do not have the right to drive to extinction such a high proportion of what are, as far as we know, our only living companions in the universe.... Second, organisms are our only means of sustainability. If we want to solve the problem of how to occupy the world on a continuing basis, it will be the properties of organisms that make it possible. Organisms are the only sustainable sources of food, medicine, clothing, biomass (for energy and other purposes).... [Peter Raven and George Johnson. *Biology*. 5th ed. St. Louis, MO: William C. Brown; 1999:556]

From tropical rain forests to oceanic reefs, we depend on a variety of ecosystems for our survival in ways we are just beginning to recognize. The twenty-first century should provide the opportunity and techniques to understand and conserve these systems far better than in the past.

The Spirit of Inquiry: Some Unsolved Problems in Biology

Science is not about getting the "right answer." Instead, science is about asking the right questions, about the process of inquiry. Children make exceptionally good investigators, because inquiry is natural to them. Full of curiosity, they not only ask questions when presented with a problem but also want to try to find the answers for themselves. For those who carry over this same high level of curiosity into adulthood and who prefer discovery over being handed authoritative "truths," science in general and biology in particular present challenging areas for exploration. Like all the sciences, biology is an organized exercise of human curiosity.

In October 1977, while exploring and mapping a portion of a rift (crack) on the floor of the Pacific Ocean near the Galápagos Islands (Fig. 1.4), geologists in a small research submarine discovered a new and unexpected world. Here, at a depth of 2650 meters (1.5 miles), hot molten rock called magma rushes out of vents along the rift, meets the ice-cold (2° C) sea water, and solidifies into black lava. In the process, the surrounding water may be heated to 350° C. The water cannot boil, however, because the pressure at that depth is far too great. Because light cannot penetrate the ocean waters that far down, it is a world of perpetual night.

Given such conditions, the researchers were hardly prepared for what they saw when they turned on the submarine's powerful spotlights. Instead of a bleak, lifeless world, they were startled to find a large community of living organisms surrounding the vents (Fig. 1.5). Tube worms, some more than a meter long, waved their reddish tentacles in the sea water. Numerous shellfish, including huge clams, were scattered across the ocean floor or attached to rocks. Dandelion-like animals called siphonophores added yet another color, and scavenger crabs scurried around among them. The geologists used the word "astonishment" to describe their feelings as their spotlights revealed these organisms for the first time.

This same word—astonishment—is found in the diaries and journals of Western explorers of the eighteenth and nineteenth centuries as they described their journeys to distant regions and saw varieties of plant and animal species quite different from those of their homelands. The vent hydrothermal community presented a similar discovery and provided biologists with a new respect for the diversity and adaptability of life. Yet the vent community also presented a series of questions: Where do its organ-

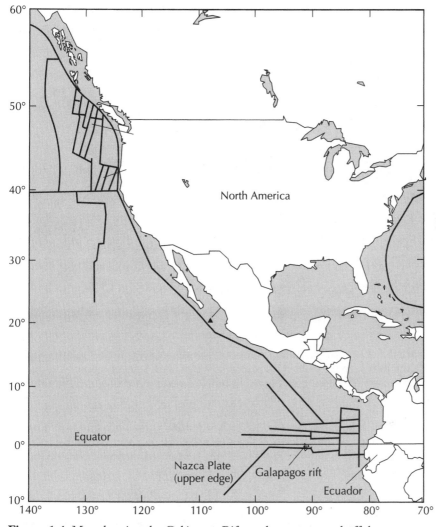

Figure 1.4 Map showing the Galápagos Rift on the equator and off the west coast of Ecuador. The rift is characterized by the emission of molten magma and hot gases from the earth's mantle into the ocean water. Except in the area at the rift center, waters at a depth greater than 1 mile are extremely cold (2° C or 35° F) and perpetually dark.

isms obtain their nutrients? How do they withstand the incredible pressure at such depths? How many millions of years have they been living in these harsh conditions, and from what ancestral forms did they evolve? Just at a time when biologists believed they had encountered all the major ecosys-

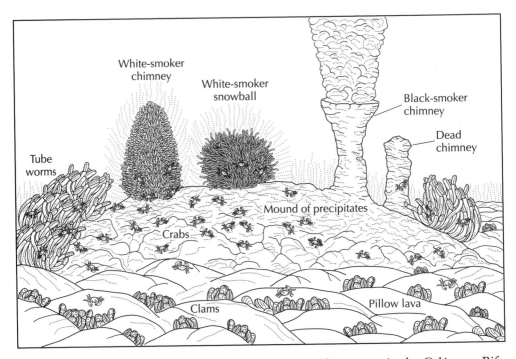

Figure 1.5 Detail of community life surrounding a vent in the Galápagos Rift, showing clams, crabs, and tube worms. An active black smoker chimney, shown to the upper left of center, puts out extremely hot water (350° C) containing hydrogen sulfide. White smoker chimneys, to the right of center, are built up out of burrows made by a little-understood organism, Pompeii worms, and emit water at a slightly lower temperature (300° C). The entire community lives on energy extracted from the breakdown of hydrogen sulfide by chemosynthetic bacteria. There is thus no energy from the sun involved in maintaining this remarkable community.

tems on earth, the hydrothermal vent communities provided a startling and unpredicted new one.

This is what the sciences in general, and biology in particular, are all about: investigation and discovery. No one can know what new phenomena investigators will encounter, and no one can anticipate the questions that new observations will raise. One thing is certain, however: *Scientific research always raises more questions than it answers*. In the case of the hydrothermal vent community, answers to some of the immediate questions were forthcoming. Additional study showed that the tube worms, clams, and several other animals feed by filtering water rich in bacteria and hydrogen sulfide (H_2S), an energy-rich compound produced by geothermal energy within the earth's interior and released in the vent's emissions. Inside the animals' body, the bacteria live in association with certain specialized tissues of the animals or sometimes even within particular cell types. Not only do the bacteria not harm their host, they actually provide the animal with basic nutrients by breaking down the hydrogen sulfide and releasing the energy it contains. This energy, in turn, is used by the bacteria to synthesize nutrients on which both the bacteria and its host depend. By the very act of this discovery, new questions emerge. The next question for investigators to pursue was exactly how the thermal energy is coupled to the synthesis of the major nutrients in bacterial metabolism.

In the remainder of this section, we examine briefly three yet unsolved mysteries of modern biology as a way to emphasize how much of science is based on inquiry and the pursuit of creative thinking.

Butterfly Migration

Many organisms regularly travel hundreds or even thousands of kilometers to areas where they can feed and/or reproduce, a process known as migration. One of the most spectacular of such long-distance migrations is that of the monarch butterfly (Fig. 1.6). This insect's range includes most of North America and part of southern Canada; its winter range is central Mexico. During the early summer weeks, the adult monarchs mate, lay their eggs on milkweed plants, and soon die. The eggs hatch into caterpillars, feeding on the milkweed plants and multiplying their original weight 2700 times in 2 weeks. Every caterpillar sheds its outer covering approximately six times during this growth process and then enters a relatively dormant stage (pupa). During this stage, the internal organs of the caterpillar degenerate and are replaced by those of the adult butterfly. In about 2 more weeks, the case containing the organism (chrysalis) opens and the adult emerges.

Newly-hatched monarch butterflies attain adulthood in early to mid-summer and reproduce, completing the cycle. Their offspring, who become adults in late summer, remain temporarily infertile. As winter approaches and the daylight hours shorten, these late-hatching monarchs begin a migration southward. This rather simple act was described in a recent popular account:

> One afternoon at the end of last August a monarch butterfly, a robust, freshly hatched male who had been cruising around for a few days in a meadow in southern Manitoba, taking nectar from asters and goldenrods, abruptly decamped and started to make his way south in a frenzy of

Figure 1.6 The monarch butterfly. © Gregg Otto/Visuals Unlimited.

flapping. He was following a migratory urge and a specific flight plan that must have been inscribed in the genes of monarchs since well before the appearance of humans. [From Alex Shumatoff. The flight of the monarchs. *Vanity Fair* 1999(Nov):270.]

Despite the absence of adults from earlier generations to show them the way, monarchs return in droves to precisely the same region of Mexico where their ancestors wintered the year before (Fig. 1.7). Here, the temperature drops sufficiently low to keep the insects in a state of semidormancy but not so cold that they freeze. As the temperature begins to rise in February and March, the monarchs become active again, gathering nectar for their return flight northward. As the days grow longer, the monarchs mate and migrate *en masse* back north to the regions from which their parents came. When they reach their northern population centers, they lay their eggs and the cycle begins again. Female monarchs often deposit some of their eggs along the way during their northward migration. When the young hatch, they continue the migration, often ending up in the same population centers as their parents.

Although the monarch migration has been known for more than a century, little was understood about what triggered it or how the butterflies found their way when none of them had been on the journey before. Indeed, the location at which the butterflies finally ended their southern migration was for many years a mystery. Some scientists believed Texas,

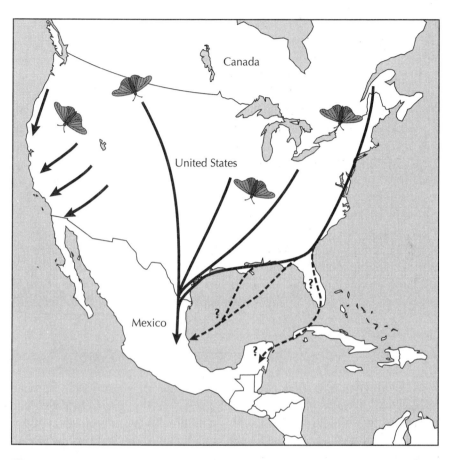

Figure 1.7 Map of North America showing the projected routes followed by monarch butterflies in autumn to their over-wintering site in Mexico. The majority of the butterflies migrating this route come from the upper midwest, but a significant population also comes from the northeast.

others northern Mexico, Costa Rica, or other parts of Central America. In the early 1970s, two lepidopterists (biologists who study butterflies and moths, which are classified in the taxonomic order Lepidoptera), Fred A. Urquart, of the University of Toronto, and the late Lincoln Brower, first at Amherst College (Amherst, MA) and later at Sweet Briar College (Sweet Briar, VA), were each interested in the monarch's migratory route. Urquart devised a system of placing a lightweight tag on the monarch's wing so that, when the butterfly was captured in the south, its origin in the north would be known. Brower, meanwhile, had been getting information from various contacts in Texas, who had spotted groups of monarchs near Big Bend National Park during the early winter. None of the groups, however, was large enough to account for the massive emigration known to take place from the north. Urquart guessed they were strays that had flown off course. More or less independently, Brown and Urquart were trying to find the butterflies' over-wintering site.

Urquart's wife, Nora, wrote an article for a Mexican newspaper in 1972, describing their search for the over-wintering site. Quite accidentally, Kenneth Brugger, an American businessman working in Mexico City, read the article and gave them a call. Although Brugger did not know one butterfly from another, he liked to drive and hike through the mountains near Mexico City. He offered to help look for the monarch site but, for several years, found nothing. Then, in 1974, he married a Mexican woman, Catalina Aguado, who accompanied him on his weekend jaunts. Being a native, she was much better than he at asking questions of local people in the mountainous areas. One day, they received a call from a 73-year-old man who said he knew a place where "the trees are filled with butterflies." It was 100 miles west of Mexico City, on rural common grounds distributed to the *campesinos* (peasant farmers) after the 1910 revolution. This land required permission from the farmers for access, a task for which Catalina proved to be an adept negotiator. The Bruggers obtained permission to visit the site and saw the massive encampment of monarchs that had eluded all previous investigators. The Bruggers immediately called the Urquarts, who came to visit the site. In their subsequent reports and publications, the Urquarts refused to divulge the exact location of the site to Brower or anyone else, on the grounds that too many visitors would disturb the fragile ecosystem that harbored the wintering insects, a decision that increased the rivalry between the two scientists.

Discovery of the site allowed Urquart to examine the monarchs and see whether he could determine the location from which they had migrated. After examining thousands of insects, he discovered one that had been tagged in Wisconsin. It was clear that this site was one of migration destinations of the North American monarchs (other sites were discovered subsequently). The site was spectacular, located on the slopes of an extinct volcano at a height of 1800 kilometers (approximately 9000 feet) elevation. At least one part of the question about monarch migration was now answered.

The other parts of the question were more intriguing: How do the monarchs locate this region, and how do their offspring find their way back to the parents' starting place in the north? What signals stimulate the monarchs to begin their migration south in late summer, or north in February and March: temperature, day length, some combination of the two, or something totally different? Urquart himself marveled:

> ...how such a fragile, wind-tossed scrap of life can have found its way (only once!) across prairies, deserts, mountain villages, even cities, to this remote pinpoint on the map of Mexico. [Quoted in Alex Shoumatoff. The flight of the monarchs. *Vanity Fair*.1999(Nov):295].

The mystery of monarch migration illustrates several important aspects of the wider nature of scientific work. Rivalries are common in science, yet they are double-edged swords. They may spur scientists on to work harder and faster than they otherwise might. However, they also may slow research when two rival researchers impede each other's efforts, for example, by withholding information. Although Brower did eventually find the monarchs' over-wintering site on his own (much to Urquart's displeasure), it took a duplication of effort and resulted in an even smaller chance that cooperative research between the two scientists would ever occur. This case also shows how important it is, especially in field studies, to enlist the help of local residents to gain access to research sites and other information. A final feature of this story is the incredible passion, as exhibited by both Urquart and Brower, that often underlies work on a scientific problem. Passion helps keep investigators motivated even when progress is slow. It also drives their persistence to stay with a problem over extended periods of time. All of these factors were involved in solving at least one part of the mystery of monarch butterfly migrations.

Monarch butterflies once again are in the news, this time in connection with a controversy over genetically engineered foods. In May 1999, the British journal *Nature* reported on a Cornell University study of monarch larvae fed on milkweed coated with pollen from a genetically engineered variety of corn containing the Bt gene (*Bt*). Researchers found that these larvae grew more slowly and had a higher mortality rate than monarchs fed on uncoated milkweed. Although *Bt* toxin has been known as a natural insecticide for years, what is new in the recent findings is that the product of the actual *Bt* gene, when ingested, is also a lethal insecticide. *Bt* corn is grown from a transgenic seed produced by several genetic engineering firms (Novartis, Monsanto, and AgrEwo) and is now being planted widely throughout the United States. The genetically engineered corn contains a gene from the bacterium *Bacillus thuringiensis* (hence the name *Bt*), which kills the larvae of a moth, the European corn borer. The *Bt* gene also has been found to kill or harm other lepidopterans, including the monarch. The implication of the Cornell finding is that, although genetically modified plants have been claimed to be ecologically safe by the biotech firms producing them, little actual data is available to evaluate what effects such new forms might have on various other species.

Hens' Teeth and Evolution

You have probably heard the phrase "scarce as hens' teeth," a wry reference to the fact that hens and, indeed, all birds have no teeth. This phrase seems at odds with the conclusion reached long ago by evolutionary biologists that birds and reptiles share a common ancestry.

Two possible evolutionary scenarios would be consistent with the idea of a common ancestor: (1) The common ancestor had teeth, but the evolutionary line leading to birds lost them long ago; or (2) the common ancestor had no teeth, and the evolutionary line to reptiles acquired them along the way. An ingenious way to evaluate these two possibilities was devised in 1980 by biologists E. J. Kollar and C. Fisher, working at the University of Connecticut School of Medicine (Farmington, CT). Kollar and Fisher took the outer tissue layers of a chick embryo, in the region that normally produces the jaws, and grew it in a special culture with the middle tissue layers from a mouse embryo, taken from the area where teeth normally develop (Fig. 1.8). In the normal embryonic development of the mouse, the enamel portion of the teeth develops from the outer tissue layer (the ectoderm), and the underlying part of the tooth, the dentin, develops from the middle tissue layer (the mesoderm). Moreover, in the mouse embryo, the outer layer does not form enamel unless it comes in contact with the underlying middle layer, which it normally does during mouse embryonic development. At the same time, the middle layer cannot form dentin unless it remains in contact with the newly differentiated enamel region.

The question Kollar and Fisher asked was: What structures might be expected to develop if chick ectoderm and mouse mesoderm were allowed to develop together in a laboratory culture? Could the chick ectoderm produce a tooth, or would it produce only normal chick mouth parts? The experiment would provide a unique way to distinguish between the two evolutionary scenarios. If the chick ectoderm could be induced by mouse mesoderm to produce teeth, then the investigators would know that the common ancestor to birds and reptiles possessed teeth and that birds lost them in their divergence. If the chick tissue could not be induced to develop teeth, then it would be likely, but not certain, that the common ancestor lacked teeth or at least that teeth had been lost very early in bird evolution.

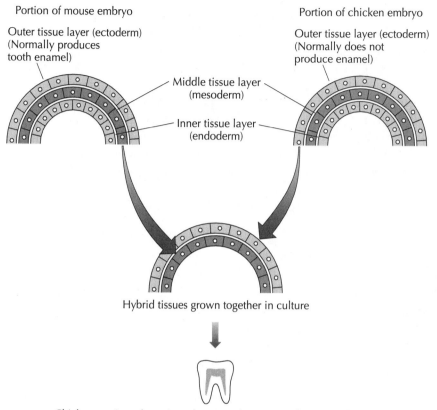

Portion of mouse embryo

Portion of chicken embryo

Outer tissue layer (ectoderm)
(Normally produces
tooth enamel)

Outer tissue layer (ectoderm)
(Normally does not
produce enamel)

Middle tissue layer
(mesoderm)

Inner tissue layer
(endoderm)

Hybrid tissues grown together in culture

Chick outer tissue layer (ectoderm) produces enamel portion of tooth

Figure 1.8 Diagram of the experiment by Kollar and Fisher in which they took tissue from the outer layer of a chick embryo (right) and grew it in culture with the middle tissue layer of a mouse embryo (left). The chick tissue, which normally does not produce teeth, was capable of generating the enamel and crown portion of a full-formed tooth in the presence of tooth-inducing middle tissue material from the mouse. Thus the chick's tooth-forming genes, latent for around 60 million years, were induced to express themselves by the presence of the appropriate tooth-inducing middle layer from the mouse embryo.

When Kollar and Fisher performed the experiment, the results were startling: The chick ectoderm did indeed develop enamel and even the crown shape of a fully formed tooth, and the mesoderm formed the dentin and the rest of the tooth (see Website 6). The evolutionary implication seems clear: The chick ectoderm, which had not produced tooth structures for 60 million years (the time at which the fossil record indicates divergence from a common ancestor probably occurred), still contained the genetic information to produce teeth. All it needed was the right environmental stimulus, in this case, teeth-inducing tissue from the mouse. The chick's own mesoderm obviously had lost the ability to stimulate enamel formation, so that, in the normal course of development, the pathway for producing teeth was blocked. Also interesting was the fact that chick ectoderm, once induced to form enamel, still retained the ability to induce dentin formation in mouse mesoderm. Even more startling, the full tooth formed by the interaction of these two tissues was not a typical mouse tooth. Although we cannot be certain, the atypical appearance might very well represent an outline of the type of tooth found in a common, long-extinct reptilian ancestor of birds.

Like all scientific investigations, the work of Kollar and Fisher raises as many or more questions than it answers. One of the most intriguing of these unanswered questions is: What keeps the teeth-forming gene or genes from expressing themselves in chickens and birds? How many other ancestral structures, not only in birds but in other animals including ourselves, lie hidden and never expressed? Why should genes that have been latent for so many millions of years still be present in birds? The fact that fossils of very primitive birds show distinct teeth (see, for example, Chapter 5, Fig. 5.4) suggests that the common ancestor probably had teeth. Why did the line developing toward birds lose their teeth? These and many other questions will provide biologists with research projects for many years to come. Research in biology, as in all the sciences, is a never-ending process.

The Origin of Life

On the last page of *On the Origin of Species*, Darwin made his only published remarks about the origin of life when, in reference to the theory of evolution by natural selection, he wrote:

> There is a grandeur in this view of life, with its several powers, having been originally breathed into a few forms or into one; and that, whilst this planet has gone cycling on according to the fixed law of gravity, from so simple a beginning endless forms most beautiful and most wonderful have been, and are being, evolved. [From Charles Darwin. *On the Origin of Species*. London, UK: John Murray; 1859:490.]

This was Darwin's deft attempt to side-step the crucial question of how the first forms of life may have arisen—the question of the origin of life itself. In a private letter written in 1871, Darwin suggested that the process might have occurred by purely chemical means "in a warm little tide pool," but he never published that idea. This was a strategic, political decision on his part, because he knew the idea of evolution would be controversial enough in its own right and did not need to be complicated by the addition of another, even more controversial, idea.

Many of Darwin's contemporaries, however, immediately raised the question of how life could have originated. The flurry of controversies in the 1860s through the 1890s on "spontaneous generation" reflected the strong interest in this age-old question. Among the many claims, for example, were the assertions that maggots, the larval form of flies, were formed spontaneously from decaying meat and that bacteria could be formed directly from organic matter (see Chapter 2). Certainly, if living organisms evolved by natural processes, then it was reasonable to argue that life might have originated by natural processes as well.

Two quite different explanations have dominated thinking about how life might have come to exist on earth: *abiogenesis*, that is, some sort of chemical processes occurring on the earth itself, and *panspermia*, the transportation of life from somewhere else in the universe. Both views have had serious proponents since the late nineteenth century and both are still hotly debated today. In the twentieth century, abiogenesis had more proponents, including renowned biochemist and population geneticist J.B.S. Haldane (1892–1961) in the 1930s and Russian biochemist A.I. Oparin (1894–1980) in the 1940s and 1950s. Panspermia has had some notable proponents, including Svante Arrhenius (1859–1927), one of the early Nobel laureates in chemistry (1908), and, currently, by Francis Crick, another Nobel laureate (1962). Today, although it is recognized that the two theories are not mutually exclusive, both are considered to be highly controversial.

The central question of abiogenesis is: Could the simple organic building blocks found in living organisms today have been produced by purely chemical means on the early earth? One of the first experiments designed to answer this question was carried out in 1952 by Stanley Miller, a young graduate student working in the laboratory of geologist Harold Urey (1893–1981) at the University of Chicago. From geochemical evidence, Urey and Miller hypothesized that the earth's early atmosphere was a chemically reducing one, that is, had little or no oxygen in it and consisted of compounds, such as ammonia (NH_3), methane (CH_4), water vapor (H_2O), and molecular hydrogen (H_2), that reduced molecules (chemically, the process of reduction involves a gain of electrons). Miller set up an enclosed, circulating gas system into which he introduced ammonia, methane, and hydrogen, along with water, all of which he kept boiling in a flask at the bottom of the tubes (Fig. 1.9). Two electrodes produced periodic electrical discharges into a large bulbous chamber at the top of the apparatus. After a week, Miller collected and analyzed the residue from the lower arm of the tube. To his amazement, he found several amino acids (the building blocks of proteins), carbohydrates (mostly simple three-carbon sugars), and several other organic compounds, including urea (a component of urine, produced by mammals as the breakdown product of proteins) and acetic acid (the central component of vinegar, produced by bacterial metabolism). After allowing the materials in the system to circulate for several more weeks, Miller was able to accumulate more than a dozen different compounds associated with living organisms. Other investigators, using similar methods but with different energy sources or slightly different starting reactants, produced a variety of other simple organic compounds. Such experiments supported the view that, on the primitive earth, many of the basic molecular building blocks of life could have been formed abiotically, that is, without already-existing life forms. Subsequent work by Sidney Fox at the University of Florida showed that, once these basic building blocks were formed, they could join together spontaneously to form larger compounds, such as simple proteins, carbohydrates, and nucleic acids.

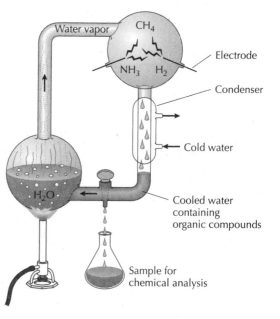

Figure 1.9 Apparatus designed by Stanley Miller for the circulation of methane, ammonia, water vapor, and hydrogen. Water was boiled in the flask at lower left. The products of the various chemical reactions that occurred were collected once a week in the trap shown at center. Energy for chemical reactions came from both heat and electric discharges in the large bulb at the upper right.

These results were exciting, but critics raised objections almost immediately. Some geochemists have questioned the assumption that the early atmosphere of the earth was a chemically reducing one. Debates about the composition of the early atmosphere are not mere quibbles over detail. Because Miller's original experiment was based on the assumption of a reducing atmosphere, his results would have been very different if oxygen had been included within his original mixture of gases. This is because oxygen, although necessary for life as we know it today, is actually quite destructive to many of the macromolecules found in living cells. If oxygen had been prevalent in large quantities in the atmosphere of early earth (it makes up 18% of the atmosphere today), it is likely that even the simplest organic molecules formed spontaneously would have been quickly broken down (oxidized). Other factors might have added to the problem, such as the turbulent conditions on the earth's surface, its excessive heat, and emission of gases such as hydrogen sulfide from volcanic eruptions. Lack of an ozone layer, which normally filters out high-energy radiation such as ultraviolet light, would have further contributed to the rapid degradation of any complex molecules that might have formed.

The original idea of panspermia was that some sorts of primitive microbes or even simpler living spores might have been transported from elsewhere in the universe by way of meteorites or other cosmic "debris." This version of panspermia has been criticized severely because of the unlikely chance that a living, organized cell or spore could withstand a thousand- or million-year journey through space, where the temperature is near absolute zero (–273° C) and there is no protection from high-energy radiation. In addition, as the meteorite containing the spores came into the earth's gravitational field and plummeted at thousands of miles per hour through the atmosphere, the high temperatures generated by the friction would have incinerated any living thing. It thus seemed unlikely that life could have begun on earth by the transport here of highly organized living structures from elsewhere in the universe.

Recently, however, an alternative to the old panspermia idea has been raised by several scientists, including Max Bernstein, Scott Sanford, and Louis Allamandola at the National Aeronautics and Space Administration (NASA) Ames Research Center (Moffett Field, CA). These investigators asked the following question: Could organic molecules, rather than fully formed spores or more highly organized living entities, have been transported to earth after the surface had cooled and at a time when those more complex molecules could have remained intact? The question could be broken down into two subsidiaries:

1. Are organic molecules detectable in various extraterrestrial sources? and

2. Is there any evidence that such molecules can enter the earth's atmosphere and reach the surface intact?

To answer the first question, the NASA team collected data from two sources: intergalactic clouds and material entering the earth's atmosphere at its outer reaches. To determine the composition of intergalactic clouds, the researchers analyzed light that passed through these clouds from more distant stars. They found that the wavelengths of light that were filtered out were similar to those observed in laboratory analyses of "clouds" containing methane, cyclic hydrocarbons, ammonia, and water. Thus, it is apparent that there probably would have been a considerable supply of the compounds necessary for life to evolve available for transport from outer space.

To answer the second question, the NASA team analyzed matter that had actually entered the earth's atmosphere from space, either as "cosmic dust" or meteorites. Here, too, the results were striking. Modern meteorites contain a variety of organic molecules, including amino acids, carboxylic acid, and ketones. Furthermore, microscopic dust particles from space collected by a research aircraft able to fly higher than 60 miles above the earth's surface contain more carbon compounds than any other interstellar materials yet analyzed. The team estimated that some 30 tons of interstellar "dust," some particles of which are no larger than a grain of sand, filter down through the earth's atmosphere every day. These meteorites and dust particles could have provided a rich source of organic matter from which more complex organic molecules on the primitive earth might have been synthesized. Thus, the new panspermia hypothesis is not so much about the transport of already living organisms to the earth but instead about what raw materials might have been available from which life could have originated over many millions of years.

As interesting as this new work may be, it raises a whole host of new questions that may be even harder to answer:

1. How could these simpler organic compounds actually join together to form larger, more complex, and varied molecules, and how did they remain intact?

2. How did these more complex molecules eventually form higher-level associations having the properties, however primitive, of modern-day cells?

3. Finally, the greatest riddle of all: How could these cells incorporate into their complex structures a process of self-replication?

These and many other questions form the basis for the ongoing study of how life originated on earth. More than either of the two preceding examples, the origin of life presents far more unanswered questions than answers. The field is only in its infancy. Yet it remains one of the most challenging and imaginative fields for investigation in the twenty-first century.

Conclusion

Enigmatic and controversial American poet Gertrude Stein (1874–1946) was one of the most innovative and acclaimed *avant garde* poets of the "lost generation" of the 1920s and 1930s. She had an early interest in science, as a graduate of Radcliffe College, an embryology student at the famous Marine Biological Laboratory in Woods Hole, and a medical student at The Johns Hopkins University. In 1946, as she lay gravely ill and sedated for surgery, her mind wandered. She murmured to her long-time companion, Alice B. Toklas, "What is the answer?" Alice, thinking her friend to be delirious, did not reply. Gertrude, always ready to fill a silence with a wry retort, answered herself, "In that case, what is the question?"

This reply encapsulates the most basic aspect of science as well as any other form of human inquiry. Inquiry is about questions. Answers are certainly important—there is no need to ask questions if we do not care about the answers. Yet the enduring part of inquiry is the way we formulate questions. The best questions in science are those that are limited in scope and thus can be answered by making specific observations or experiments. Like everything else, there is a process involved not only in answering but also in asking questions. It is in the asking of the questions that knowledge begins.

Further Reading

On the growth of biological thought (nineteenth and twentieth centuries)
Allen, Garland E. *Life Science in the Twentieth Century*. New York, NY: Cambridge University Press; 1978.

> *Chapters 1–4 specifically detail the change in the practice of biology from a descriptive to an experimental science.*

Mayr, Ernst. *The Growth of Biological Thought*. Cambridge, MA: Harvard University Press; 1982.

> *A long and detailed book that explores the nature of evolutionary theory as an example of the shift from "old" to "new" methods in biology.*

On the general topic of questions in science

Goldstein, Martin, and Inge F. Goldstein. *How We Know. An Exploration of the Scientific Process.* New York, NY: Plenum Press; 1978.

> *Chapters 1 and 2, in particular, are concerned with the general process of science, including asking questions.*

On the discovery of the Galápagos vent communities

Cone, Joseph. *Fire Under the Sea: The Discovery of the Most Extraordinary Environment on Earth: Volcanic Hot Springs on the Ocean Floor.* New York, NY: Morrow; 1991.

> *A useful source for understanding the geology, ecology, and organismic physiology of the vent community.*

MacDonald, Ken C., and Bruce P. Luyendyk. The crest of the east Pacific rise. *Scientific American* 1981;244(May):100–116.

> *A discussion of both the geology of the ocean rift areas and the biology of rift communities. Relates the formation of rifts and geothermal activity to the geological theory of plate tectonics.*

On monarch butterfly migration

Larsen, Torben. Butterfly mass transit. *Natural History* 1993;102(June):31–38.

> *Describes migrations of butterfly species in parts of the world other than North America, particularly the Asian subcontinent and Africa.*

Shoumatoff, Alex. "The flight of the monarchs. *Vanity Fair* 1999;(Nov):268–300.

> *A well-written popular account of the monarch butterfly problem, including a short discussion of* Bt *corn.*

Urquhart, Fred A. *The Monarch Butterfly: International Traveler.* Chicago, IL: Nelson Hall; 1987.

> *An excellent book discussing natural history, physiology, behavior, and experimental work on monarch butterfly migration. Contains many useful illustrations.*

On "hens' teeth"

Gould, Stephen J. *Hen's Teeth and Horse's Toes.* New York, NY: W.W. Norton; 1983.

> *Contains a number of interesting essays on evolutionary theory, including one for which the book is titled, about the reactivation of long-lost ancestral traits in modern organisms.*

On the origin of life

Bernstein, Max P., Scott A. Sandford, and Louis J. Allamandola. Life's far-flung raw materials. *Scientific American* 1999;281(July):42–49.

> *A clear and comprehensive discussion of the modern panspermia idea that organic material and water were transported to Earth from intergalactic clouds and dust.*

Schopf, J. William. *Cradle of Life. The Discovery of the Earth's Earliest Fossils.* Princeton, NJ: Princeton University Press; 1999.

Although mostly about microfossils in the pre-Cambrian era (roughly 1–4.5 billion years ago), this book contains several good chapters (4 and 5) summarizing current thought about abiogenesis.

Zubay, Geoffrey. *Origins of Life on the Earth and in the Cosmos.* 2nd ed. New York, NY: Academic Press; 2000.

This textbook contains much useful reference material on aspects of the origin of life, including abiogenesis and modern versions of panspermia. Good introductory chapters on the origin of the universe, the elements, our galaxy and solar system, and the evolution of the atmosphere of early earth.

Suggested Websites

1. MendelWeb

http://www.netspace.org/MendelWeb/

Centered around an extensively annotated copy of Gregor Mendel's 1865 paper "Experiments in Plant Hybridization" (in English and German), MendelWeb allows students to study Mendel's paper and, in turn, learn basic information about introductory data analysis, elementary plant science, and the history and literature of science.

2. The Human Genome Project

http://www.ornl.gov/TechResources/Human_Genome/home.html

An up-to-date site maintained by the Department of Energy concerning the scientific aspects, medical uses and ethical, legal, and social issues surrounding the Human Genome Project.

3. Black Smokers

http://www.amnh.org/nationalcenter/expeditions/blacksmokers/glossary.html

This site chronicles the summer 1998 expedition of a group scientists, engineers, and educators from the American Museum of Natural History Museum as they collected black smoker sulfide chimneys from the San Juan de Fuca Ridge. Links provide basic information about hydrothermal vents and vent communities and access to the scientists involved.

4. Monarch Migration

http://www.learner.org/jnorth/spring2001/species/monarch/index.html

A weekly chronicle of the monarch's migration to and from Mexico. A yearly event, previous years' activities are archived at this site. This is part of a more extensive site concerning animal migration.

5. Bt and Monarchs

http://www.biotech-info.net/butterflies_btcorn.html
Part of a larger "Ag BioTech InfoNet" site that providing "genetic engineering applications, impacts and implications," this page is full of links to valuable information about the effects of Bt on monarch butterflies.

6. Hen's Teeth

http://zygote.swarthmore.edu/prox1.html
Observe the actual tooth produced by Kollar and Fisher's amazing experiment.

7. Origin of Life, Exobiology, Astrobiology: a page of links and more

http://web99.arc.nasa.gov/~astrochm/originlinks.html
An excellent sampling of available links on the origin of life and related subjects compiled by NASA.

The Nature and Logic of Science

In the spring of 1985, a reporter for the British Broadcasting Company (BBC) interviewed a biologist on the fertilization of human eggs and development of human embryos outside the mothers' bodies. "Surely," the reporter asked, "you must have done a lot of research to get this far?" The biologist nodded affirmatively. "Well, then," the interviewer asked, "don't you know everything you need to know?"

The biologist was at a loss for words. It was a question no scientist would think of asking and was based on several widespread misconceptions about the nature of science. One such misconception is that science is largely a matter of collecting facts into a sort of encyclopedia of knowledge about a particular field, for example, biology, chemistry, or physics. Related to this misconception is the notion that the number of such facts is finite, and, theoretically at least, we can expect to discover all of them if we search long enough. Yet it is doubtful that the reporter would have asked a historian if he or she knew enough about British or Russian history or a poet whether he or she had written enough poems. Science writer Ted Neild ended a critique of this BBC interview with the following observation:

> Science is about ideas...and, because...the unifying ideas [of science]...are in constant need of revision as new facts come to light, and the insight of new ideas redefines the implications of the old knowledge, we can never 'know enough.' This is so obvious to scientists, yet apparently obscure to everyone else. [*New Scientist* 1985(March 7):46.]

What, then, *is* science, and why is it so often misunderstood? These questions lead us into the main topic of this chapter: the nature and logic of science.

Science as an Intellectual Discipline

Defining Science

The term "science" was first coined in 1851 by British philosopher William Whewell (1794–1866) to refer to the study of the natural world. The term comes from the Latin *scientia*, meaning "knowledge" or "knowing." Before Whewell's time, people who studied nature were called natural philosophers, a term emphasizing the close historical connection once existing between the sciences and the humanities.

In more recent times, however, science has become an increasingly separate pursuit. Yet definitions clearly distinguishing science from other fields of human thought are difficult to formulate. This is partly because science is not a single enterprise but a collection of activities that often varies widely from one scientific discipline to the next. Moreover, none of the methods associated with science are unique to science. A paleontologist who studies fossils to reconstruct their evolutionary history is likely to have more in common with a historian than with those biologists who carry out laboratory research in fields such as genetics or physiology. Both paleontologists and historians attempt to reconstruct the history of events that occurred in the past, and both must draw conclusions from the only evidence available to them. By contrast, biologists studying processes taking place in organisms that are alive today have available to them a variety of methods not available to paleontologists or historians, most especially the use of experimentation. Despite these differences, few today would seriously question that paleontology is a science.

Characteristics of Science

Although we may not be able to define science as a unique activity, we can describe it as a specific combination of shared practices and assumptions. All science is based on empirical knowledge that is, knowledge obtained through our senses of sight, sound, touch, taste, or smell. Empirical knowledge, even of the most tentative or casual sort (for example, observing a moth emerge from a cocoon or the sun rising), is often the starting point for scientific investigation.

A second characteristic of science is its commitment to rationality. Rationality involves seeking explanations in terms of natural causes. Rational explanations are those that can be understood in terms of natural as opposed to supernatural processes. No astronomer today is satisfied with the medieval explanation that planets are moved through their orbits by angels, and no biologist is satisfied with interpreting disease as the result of divine punishment.

A third characteristic of science is its emphasis on repeatability. Scientific results are always subject to confirmation by other investigators. Along with repeatability goes another related characteristic: reliability. Numerous persons claim to have seen the Loch Ness monster, a prehistoric reptile purported to live in a very deep and ancient lake in northwestern Scotland. Such claims have never convinced biologists that "Nessie" really exists. The observations lack reliable confirmation and thus are suspect.

A fourth characteristic of science is its commitment to testability. No matter how interesting or imaginative an explanation might be, it is of no value if it cannot be tested. Ideas that cannot be tested provide no basis for asking further questions or for carrying out additional research. As fascinating as they might be, such ideas are intellectual dead ends in science.

Following directly out of this fourth characteristic of science is a fifth: its commitment to the use, whenever possible, of experimentation. An experiment is a planned intervention into a natural process to observe the effects of that intervention. For example, a biologist interested in how early stages in the development of frog embryos affect the final form of the adult may remove structures from embryos of different ages and record changes occurring in the adult. Such experiments allow the investigator to ask specific questions and obtain equally specific answers. They also allow him or her to make observations that might never be made under natural conditions. Of course, not all fields of science lend themselves to experimentation. The film *Jurassic Park* notwithstanding, evolutionary biologists cannot step back in time and experiment with dinosaurs, nor can astrophysicists experiment with stellar systems. Yet even in those fields where experimentation is less applicable, it remains a goal whenever possible.

A sixth characteristic of science is its search for generality, the establishment of general principles operating in the natural world regardless of differences in time or place. Physicists are interested in principles of motion or gravity that apply not only at different places on earth but throughout the universe. Biologists are interested in understanding living processes in all organisms, not merely mice, maple trees, or bacteria. Although it is true that each type of organism will respond to its own unique set of environmental conditions and its own internal make-up, the same basic biological principles are believed to apply throughout the living world.

The Relationship Between the Sciences, Social Sciences, and Humanities

The six characteristics listed in the previous section are not necessarily unique to science. Historians, writers, painters, musicians, and others often try to build more general concepts from specific and precise observations. (For example, many critics believe that the best poets use the most concrete imagery.) All try to formulate these concepts in universally understandable terms. In the seventeenth century, for example, Isaac Newton (1642–1727) formulated a concept of the universe as operating by mechanical principles. The solar system was envisioned in terms of a common machine of the day, the clock. Indeed, the phrase "clockwork of the heavens" was commonly used to refer to the Newtonian universe. Science has as much room for creativity and imagination as do any of the arts. Creative efforts in science are like a painter's canvas that is constantly being reworked to obtain greater accuracy, completeness, or aesthetic appeal.

The sciences, social sciences, and humanities have long interacted in ways that lend support to one another. Questions about the use of atomic energy, the application of our knowledge of biology to preserve the environment, or the ethical implications of new discoveries in genetics often make us keenly aware of some of the ways in which science inspires our social and humanistic concerns and vice versa. Science has inspired many poets and creative writers. (The vast literature falling under the heading of science fiction is but one of many examples.)

The reverse is also true, however. Social, political, philosophical, and even artistic developments often interact with and directly affect the course of scientific discovery. Historian Samuel Y. Egerton has argued that studies in visual perspective by early Renaissance painters, such as Giotto de Bondone (1337?–1376?), initiated precise, quantitative, and mechanical methods of viewing nature that served as crucial stimuli for the scientific revolution in astronomy and physics in the sixteenth and seventeenth centuries. The resulting view, that of an infinite universe working by purely mechanical principles, greatly altered human beings' views of themselves. Historian Robert Young has argued that if Darwin had not been familiar with the works on political economy of Adam Smith (1733–1780), David

Ricardo (1773–1833), and Thomas Robert Malthus (1766–1834), all of whom emphasized the scarcity of resources and the competition for those resources as a natural part of economics, he quite possibly never would have come up with the idea of evolution by natural selection. Borrowing these ideas from the social and economic realm was one of Darwin's most creative efforts. For their part, Darwin's ideas stimulated another profound revolution in the way human beings viewed themselves as having evolved like other organisms from previously existing forms. Thus, our social views may influence the way we form our picture of the natural world as much as our picture of the natural world may influence our social views.

Science, Common Sense, and Intuition

A key feature of scientific thinking is that it tends to be highly suspicious of intuitively derived decisions lacking empirical backing. No matter how intuitively obvious an answer might seem, it must always be analyzed critically.

Yet intuition can be an important part of the scientific process. Such famous scientific ideas as Einstein's theory of relativity and the molecular structure of the gene were, to a considerable degree, the results of intuition. However, neither these nor any other scientific concepts would have survived for long had they not been subject to empirical verification. It is no less so for much of ordinary human experience. Although intuition may be highly useful both in analyzing problems and generating ideas, *in science it must always be confirmed or rejected empirically*.

The Logic of Science

The elements of logic are at the heart of the methods employed in all rational thinking. To understand how we think logically, an examination of the components of empirical knowledge—observations, facts, and conceptualizations—is in order.

Observation and Fact

The foundation of all empirical knowledge is a discrete item of sensory data. The statements "the car is red" or "this bird's song consists of three notes" represent observations. Each encompasses a single item of sensory data, in these cases, sight and sound, respectively. In science, investigators often employ various instruments to make observations that none of our senses is able to detect. Microscopes, for example, magnify objects too small to be seen with the unaided eye, high-frequency audio detectors pick up and record sounds that our ears cannot hear, and electronic sensors amplify subtle chemical changes in living tissue that none of our senses can detect. Of course, as any trial lawyer will attest, two observers do not always see the same object or event in the same way. A color-blind person may see a red car as a shade of gray, or someone who is tone-deaf may hear a bird's song as a single tone rather than as three distinct notes. For observations to form the basis of our ideas, they must be agreed upon by different observers. The criterion of repeatability discussed earlier means that any and all observations must be checked not only by the original observer but by other observers as well.

As they are repeated and agreed upon by some community of observers, observations often become established as facts. Facts, then, may be defined as individual observations that have been confirmed and accepted by consensus. For the group involved in that consensus, at least, the observations have been established as facts. It becomes a fact that the car is red or the bird's song consists of three notes *because and only because a group of observers has agreed it is so.* Facts are not some inalterable truth handed to us by an impersonal natural world but instead are negotiated agreements among individuals as they compare their observations.

Does the preceding mean that what we term "facts" are arbitrary? After all, if a color-blind person sees a car as gray, is that not a fact for that person, even if other people see it as red? The answer is yes in one sense but no in another. Because knowledge is gathered and becomes useful *only* in a social context, communication and agreement about what is and is not fact is critical in verifying observations and establishing facts. For example, a small group of Elvis Presley enthusiasts have claimed to have seen the singer alive and that he has spoken to them. For this group, it is a fact that Elvis lives. Similarly, for many people, the existence of unidentified flying objects, or UFOs, is also a fact. As more and more observers' views are included, however, the consensus dwindles and such former facts eventually may be regarded by the majority as unsubstantiated. This is where the empirical component of scientific knowledge becomes important in deciding what will or will not be accepted as fact. No matter how many people claim that something is a fact, if it cannot be observed repeatedly by a wider circle of observers, the "facts" become highly questionable.

One of the greatest strengths of science is its total commitment to putting observations or facts to empirical testing. It is this commitment that distinguishes science from areas such as religion or other forms of supernaturalism. To claim, as do some advocates of the supernatural, that it is necessary to be a "believer" in the phenomenon in order to be able to observe it is to beg the question, because such a claim amounts to little more than saying we can simply believe what we want to believe. One advantage of examining our thought processes is to be able to share with each other some common methods of understanding the world. Humans are social animals, and the knowledge we generate is not merely individual knowledge. We could never exchange ideas if we did not employ some common methods of drawing conclusions about the phenomena we encounter.

Recognizing the social component involved in establishing observations and the facts that we derive from them does not lessen their value but, instead, suggests that observations and facts are, to some extent, the product of a specific historical time and place. In other words, observations and the facts that derive from them often are not independent of the observer but are very much the product of humans interacting with the world and each other. Recognizing the social nature of observations and facts tells us that, if two or more people do not agree on the facts, they will have little success in discussing the conceptualizations that may be derived from those facts. It would be useless, for example, for two people to debate whether extraterrestrial UFOs come from inside or outside of our solar system if they cannot agree that UFOs exist in the first place.

From Fact to Conceptualization

Human existence would be quite chaotic if our total experience consisted only of discrete observations, even if these were all quite well established. The agreed-upon fact that the sun rose over the eastern horizon this morning would be relatively useless if we could not place it in some larger framework, or conceptualization. Conceptualizations are abstract statements that go beyond individual facts and relate them to one another. Conceptualizations may be simple generalizations: "The sun always rises on the eastern horizon." They may involve more complex explanations: "The sun rises in the morning and sets in the evening because the earth is turning on its axis." In either case, the most important characteristic of such conceptualizations is that they bring a group of specific facts together into a more general and useful relationship, allowing us to organize these facts into patterns of regularity and to make accurate predictions. Indeed, it has been said that science is the search for patterns in nature.

If it is true that conceptualizations depend upon the observations and facts at hand, it is equally true that the kinds of conceptualizations we generate determine something about the observations we make. We tend to perceive readily only that which we are prepared by our conceptualizations to see. Seeing or observing is not an automatic activity. Art teachers are fond of saying that "art teaches you to see," which means that observation is something we all have to learn about as a process. Our eyes and ears may be open, but seeing and hearing involve the brain's integrative mechanisms, which, in turn, are molded by our learning experience. Consciously or unconsciously, people often choose to overlook observations that do not fit with what they conceive to be true. The value of examining these philosophical issues is that it can make us more aware of how various factors influence the way we observe the world or put our observations together into conceptualizations. In turn, such awareness may help provide an antidote to any unintended biases that might influence our thinking.

Types of Conceptualizations

Although many categories of conceptualizations are found in science, we will limit our discussion here to three major types: generalizations, hypotheses, and theories.

Generalizations. A generalization is a statement that is meant to apply to a large class of objects or set of phenomena. The example given previously, "the sun always rises in the east," is a generalization about a daily occurrence. Similarly, the statement that "all dogs have four legs" is a generalization about a set of objects, dogs. Generalizations are based on summarizing a number of specific observations of the same processes or object. Generalizations are highly useful, because they point to a regularity among phenomena in the material world. They become problematic only when they are based on a very small number of cases that may not adequately represent the whole.

Hypotheses. Hypotheses are tentative explanations to account for observed phenomena. For example, if you flip the switch on a lamp and it does not go on, you might formulate a hypothesis proposing as an explanation that the bulb is burned out. This hypothesis leads inevitably to a prediction: Replacing the bulb should make the lamp light again. In science, hypotheses always lead to predictions that can be verified or refuted. *The formulation and testing of hypotheses lie at the very heart of any scientific or rational inquiry.*

Theories. Although philosophers of science often differ as to the precise definition of the term "theory," practicing scientists generally agree that a theory is an explanatory hypothesis that has stood the test of time and is supported well by the empirical evidence. Theories generally are broader, more inclusive statements than hypotheses and may often relate two or more hypotheses to one another. The Darwin–Wallace theory of evolution by natural selection, for example, incorporates hypotheses dealing with the modifiability of organisms by selective breeding; the meaning of similarity of structure and function; the role of competition for food, territory, and other natural resources; factors involved in mate selection; and others.

Observation, Fact, and Conceptualization: A Case Study

At the turn of the century, biologists made numerous observations to determine the number of chromosomes (from Greek *chroma*, color, and *soma*, body) present in the cells of most plants and animals, including humans. To see chromosomes, it is necessary to stain cell preparations for viewing under the microscope. Chromosomes so treated appear as dark, oblong objects surrounded by other partially stained material in the nucleus.

Figure 2.1A suggests some of the problems observers encountered in trying to make accurate chromosome counts. First, the chromosomes usually were clumped together in such a way that it was not always easy to tell where one ended and another began. Second, females appeared to possess

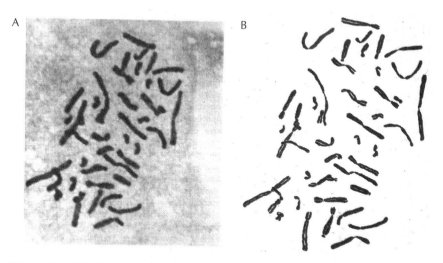

A   B

Figure 2.1 (A) Photograph showing chromosomes from a stained spleen cell culture of a 17-week human fetus. Note how the chromosomes clump, making accurate counts difficult. (B) Camera lucida drawing of the same group of chromosomes. Such drawings help to elucidate detail more clearly, because they represent a composite of observations at different depths of focus—something no single photograph could provide. [Reprinted from T.C. Hsu, Mammalian chromosome in vitro. I. "The karyotype of man." *Journal of Heredity* 1952;43:168.]

one more chromosome in each body cell than did males. Third, chromosomes usually curled and twisted toward or away from the plane of the field of vision. Consequently, it was easy to count as two chromosomes what was actually one chromosome appearing at different focal planes under the microscope.

In the early part of the twentieth century, microscopists often used a camera lucida for recording microscopic observations. The camera lucida projects what is observed under the microscope onto a flat surface and allows the observer to trace a pattern of the projected image (Fig. 2.1B). It was far easier to count chromosomes in camera lucida drawings than in the actual chromosome preparations. Yet, even then, the same observers did not always see the same number of chromosomes (Fig. 2.2). In 1907, German cytologist H. von Winniwarter made the earliest counts of human chromosomes and reported 47 chromosomes to be the total in humans: 23 pairs plus an extra "accessory" chromosome, the X chromosome. Between 1921 and 1924, however, T.S. Painter (1889–1969), then at the University of Texas, developed new techniques for preparing and observing chromosomes. Using these methods, he reported a count of 48 chromosomes, or

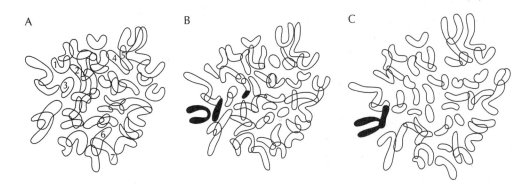

Figure 2.2 Three different camera lucida tracings of the same chromosome preparation as drawn by different sets of observers: (A) Evans, (B) von Winniwarter, and (C) Oguma. Note the difficulty in determining whether some parts of chromosomes are attached to or separate from other parts. To see how this affects actual counting, observe the chromosome labeled 2 in image A. Evans saw number 2 as a long, crescent-shaped chromosome, whereas both von Winniwarter and Oguma saw it as two shorter, separate chromosomes. Evans saw chromosome number 3 as a single chromosome, whereas von Winniwarter and Oguma saw it as two separate ones. Such problems greatly confused the issue of human chromosome counts.

24 pairs. Between 1932 and 1952, at least five other observers confirmed Painter's count of 48 chromosomes. By the early 1950s, it was accepted that the correct chromosome number for the human species was 48, and all biology textbooks dutifully gave that number.

The certainty of Painter's count had two negative effects. First, it stifled further investigation. The count became authoritative, and people simply stopped counting human chromosome preparations. Second, it prejudiced the few chromosome counts that *were* made. The conceptualization that 48 was the correct number caused observers to believe they saw 48. In the 1930s and 1940s, however, several new techniques were introduced. One of these was the preparation of karyotypes. To prepare karyotypes, cell chromosomes are photographed and the individual chromosomes cut out of the photographic print. The chromosome images are then arranged on a sheet in a systematic fashion, making it possible to account for each individual chromosome and match it with its partner (Fig. 2.3). In 1955, Eva Hansen-Melander and her colleagues in Sweden had been studying the karyotypes of cancerous human liver tissue. They consistently counted 46 chromosomes. Eventually, after some doubts about the accuracy of their own observations, Hansen-Melander's group challenged in print the long-established notion that humans have 48 chromosomes. By 1960, after many confirmations of the number 46, it was agreed that the older count of 48 was wrong.

What can we conclude about the process of observation from the history of establishing the chromosome count in humans? First, observation is not a simple, passive process in which the observer simply lets sensory data flow into his or her brain. It is an active process that involves a good deal of input or active construction on the part of the observer. We have to "learn to observe." Second, because observations depend upon sensory data, if the material being observed is itself ambiguous (such as clumped chromosomes) the observations will reflect that ambiguity in one way or another, either in disagreements among different observers or failure of a single observer to confirm earlier findings. Third, observations generally contain some subjective input. Determining whether a specific stained mass represents one or two chromosomes is often a judgment call. Fourth, *people often find what they expect to find.* The expectation that human cells contain 48 chromosomes caused many workers actually to "see" 48 chromosomes. Thus, even when something else is observed, the force of accepted dogma may cause investigators to disbelieve their own sense impressions. Fifth, it should be pointed out that Painter was working with testicular samples taken from a patient at a Texas mental hospital. This individual

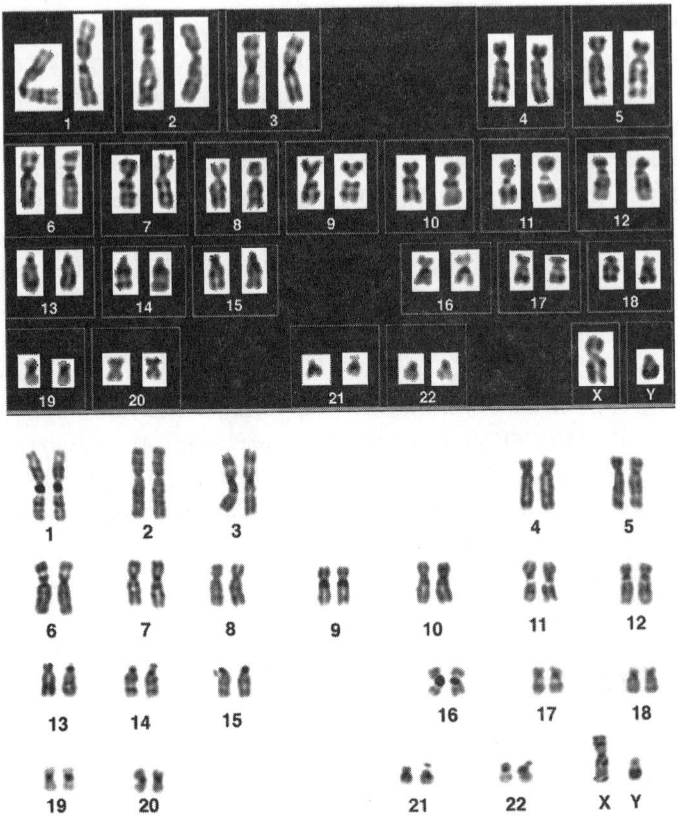

Figure 2.3 Karyotype of a human male. The chromosomes are arranged in their natural pairs by number (numbers start with the larger chromosomes). Characteristically, the 23rd pair is the sex-determining pair, shown here as XY, for a male (females would be XX). [Courtesy Dr. Thomas Ried, M.D. and Dr. Hased M. Padilla-Nash, Cytogenetics Laboratory, National Cancer Institute, NIH.]

may well have possessed an extra chromosome, a condition sometimes associated with certain types of mental retardation. Thus, Painter's observations may have been accurate but his starting material atypical. Finally, this case also shows how the introduction of a new technique, in this case karyotyping, may change a conceptualization in science by improving the accuracy of observations.

Creativity in Science

A popular stereotype maintains that creativity is a warm and lively process reserved for poets, musicians, and other artists, whereas science, by contrast, is coldly logical. In truth, science may be just as creative as any of the arts and the scientist as much a creator as the poet. When Darwin and Alfred Russel Wallace (1823–1913) independently read Thomas Robert Malthus's *Essay on Population* (1798), the concept of natural selection as a driving force in evolution occurred immediately to both of them. In each case, an act of creativity resulted in the formulation of a bold new concept. Exactly *how* each individual arrived at this concept is impossible to pinpoint. As in the artistic world, the process of creativity remains elusive. The momentary insight, the creative flash of inspiration, often happens so quickly that the individual involved may have difficulty in reconstructing the creative process. Thus, in our discussion of the nature of scientific thought, we will not be able to say much about the creative act of idea formation itself. However, this should not be taken to mean that creativity is unimportant in science. On the contrary, it often plays a central role. What we can understand more fully is the process of verification, that is, how we formulate and test hypotheses in a conscious and logical way.

The Logic of Science: Induction and Deduction

We now turn to the more formal aspects of hypothesis formulation and testing, by examining the processes of induction and deduction.

Induction and Deduction

A major pattern of thought involved in forming conceptualizations is inductive logic or induction Induction is the process of making general statements based upon a set of individual observations. Consider for example, the following series of integers: 2, 4, 6, 8, 10, 12, 14, 16, 18.... Here, individual observations might lead to the formulation of a hypothesis proposing that the entire series, including any yet-to-be-revealed numbers indicated by the dots, is composed of the positive, even integers. Or, suppose a person tastes a green apple and finds it to be sour. If the same person tastes a second, third, and fourth green apple and finds them to be sour also, he or she might reasonably hypothesize that *all* green apples are sour. The concepts that the entire number series is composed of even integers and that all green apples are sour are examples of hypothetical generalizations formed by inductive logic. They are inductive generalizations. Such inductive generalizations not only summarize a set of observations, they also may serve to provide predictions about as yet unobserved events, for example, the identity of the next number in the series or the expected taste of the next green apple.

Going beyond the formulation of hypotheses formed by induction to test whether they are correct involves the use of deductive logic, or deduction. Often called "If...then" reasoning, deductive logic is the heart and soul of mathematics (for example, "*if* two points of a line lie in a plane, *then* the line lies in the same plane). Deduction is no less important in science. The "if" portion of the if...then format represents the hypothesis. The word "if" stresses the tentative, or conditional nature of a hypothesis, whereas the word "then" (or *prediction*) stresses that the next statement (or *conclusion*) *follows inevitably from acceptance of the hypothesis.*

A deduction is said to be valid if the conclusion follows necessarily from the original hypotheses from which it is derived. We can see this more clearly by laying out the logical sequence involved in what logicians call a deductive syllogism, a formal sequence of if...then statements:

> *If* . . .   the number series consists of positive even integers,
>
> *then* . . .   the next number to appear in the number series will be 20; and
>
> *if* . . .   all green apples are sour,
>
> *then* . . .   the next green apple tasted should be sour.

Both syllogisms are valid. As long as we accept the two hypotheses as stated, *we have no choice but to accept the conclusion.* The following, however, is an invalid syllogism:

> *If* . . .   this fruit is sour,
>
> *then* . . .   it is a green apple.

Here, the conclusion does *not* follow logically, because sour fruit forms a larger set than green apples and may include many other sour fruits that are not green apples. Because the conclusion does not *necessarily* follow from the hypothesis, it must be invalid.

Note that the conclusion of a deductive syllogism is also a prediction, that is, it makes a statement about some future event. If a number series consists of positive even integers, then the number after 18 would be predicted to be 20. Similarly, if all green apples are sour, the next green apple tasted should be sour. The fact that deductive syllogisms lead to predictions means that their hypotheses can be tested. Testing hypotheses is one of the cornerstones of scientific investigation. This general method of reasoning is referred to as the hypothetico-deductive method.

Thus far we have been evaluating validity in a strictly logical sense. What about the "truth" of a statement in the real world? Although one syllogism concluded that this green apple must be sour does not mean that it *is* sour. *Validity and truth are not the same.* Validity has to do with logic, truth with our experience in the real world. For example, we can set up a perfectly valid deductive syllogism that has nothing to do with truth in terms of human experience:

> *If* . . .   all geometric figures have four corners, and
>
> *if* . . .   all circles are geometric figures,
>
> *then* . . .   all circles have four corners.

This syllogism is valid, yet, in the world of plane geometry, it clearly is not true that all circles have four corners.

Logic, Predictions, and the Testing of Hypotheses

The use of observations or experimentation to test hypotheses implies that there is a distinct relationship between hypotheses and the predictions they generate. This relationship is shown in the "truth table" in Figure 2.4.

Note first that, barring an error in carrying out the experiment itself, obtaining a false prediction automatically implies that the hypothesis also must be false, because, as the truth table shows, *a true hypothesis can never give rise to a false prediction.* This becomes obvious if we go back to our examples of the number series and sour green apples. If the next number after 18 turns out to be 19 instead of 20, the hypothesis proposing that the series consists of positive even integers is clearly false. Similarly, if the next green apple is sweet, the hypothesis proposing that all green apples are sour also must be false. Logically speaking, therefore, we must reject both hypotheses. In the real world of formulating and testing scientific hypotheses, however, seldom does a single counterexample lead to the full-scale rejection of a hypothesis, especially a widely accepted one. More likely, the "all green apples are sour" hypothesis would be modified, perhaps to "*Most* green apples are sour." There are good reasons why logic alone does not necessarily prevail here. A green apple that is sweet might be a different variety of apple that remains green when ripe; most certainly such varieties exist.

The "Truth Table"

Hypothesis	Conclusion (prediction)
True	True
False	True or false

Figure 2.4 The "truth table" shows the relation between a hypothesis and its prediction. Note that a true prediction may be derived from a false as well as a true hypothesis. Therefore, true predictions do not constitute proof for the truth of a hypothesis. False predictions, on the other hand, derive only from false hypotheses, so that falsifying a prediction leads to certainty (in the logical sense) in rejecting the hypothesis on which it is based.

The truth table also shows that *only a false hypothesis can give rise to a false prediction.* The importance of this last statement cannot be overemphasized, for it is the only instance in which we can establish absolute certainty in evaluating scientific hypotheses. Note, on the other hand, that *obtaining a true prediction cannot achieve absolute certainty about the truth of scientific hypotheses, because false hypotheses also may give rise to true predictions.*

The Concept of Proof in Science

Deductive logic is the heart and soul of mathematics. It is no less important for science. In mathematics, however, proofs by deduction are the standard. Scientific proofs, on the other hand, are a combination of induction and deduction and therefore are *never more than probable.* Indeed, the word "proof" should not be used in the context of science at all.

Let us see why this is the case. You may recall from high school algebra the proof that the square root of 2 ($\sqrt{2}$) is an irrational number, that is, cannot be expressed as a ratio of two integers, for example 1/2 or 3/4. It is possible to *prove* this, because the set of all numbers is divisible into two subsets: those that are rational (that is, *can* be expressed as a ratio of two other numbers) and those that are irrational (that is, *cannot* be so expressed). The first step in the proof involves putting forth the hypothesis that the square root of 2 *is* a rational number, that is, can be found in the set of rational numbers. The algebraic manipulations that follow lead eventually to a contradiction of this hypothesis, that is, to a false conclusion or prediction. This means that, because the square root of two is not in the set of rational numbers, it can only be in the other set, that of the irrational numbers.

This situation, which can be summed up as "if not *a*, then *b*," is not attainable in the real world of the scientist. The experimental disproof of any one scientific hypothesis does not mean that an alternative hypothesis must be true. Instead, any of a large number of alternative hypotheses might account for the phenomenon (for example, the cause of a disease) being researched. The conclusion here is an important one: *Science cannot prove anything.* Popular reporting to the contrary, science has not "proved" that cigarette smoking causes such conditions as lung cancer, emphysema, and/or heart failure. On the other hand, a vast amount of evidence supports the scientific hypothesis proposing a link between smoking and these conditions.

Although science cannot prove, it *can* disprove. This is because, as the truth table shows, failure to verify the prediction or predictions that necessarily follow from the acceptance of any hypothesis shows conclusively that, barring experimental error, the hypothesis must be false. And, as the truth table shows, only false hypotheses generate false predictions.

Many hypotheses we believe to be false today once were accepted by scientists and lay persons simply because they led to accurate predictions. For example, acceptance of the hypothesis that the sun orbits Earth (the geocentric, or earth-centered, view of the universe) rather than vice versa (the heliocentric, or sun-centered, system) leads to the prediction that the sun will rise on one horizon, cross over the sky, and set on the other horizon—and it does so every day. The fact that this prediction turns out to be correct does not, of course, mean that the sun actually *does* orbit Earth (refer back to the truth table). To demonstrate that this hypothesis is false, other test observations are necessary. One such test is to predict future relative positions of the sun, Earth, and the other planets in our solar system, based upon acceptance of the geocentric hypothesis. Such predictions are invariably false, thus allowing us to reject the hypothesis. On the other hand, the hypothesis proposing that Earth and the other planets in our solar system orbit the sun leads to much more accurate predictions about the relative positions of these bodies at any point in time. Each such accurate prediction lends support to the heliocentric hypothesis and supports rejection of the geocentric hypothesis.

The "Dissection" of an Experiment

We turn now to an early example of a scientific investigation that demonstrates the nature of the logical basis of science.

Today we know that semen, produced by male animals (including humans), contains spermatozoa, or sperm. Sperm are living cells, each of which has a headpiece and tail and conveys the inheritance factors (genes) of the male to the female ovum, or egg. In sexual reproduction, the sperm and egg unite in the process of fertilization, leading to the subsequent embryological development of the organism.

This is what we know *today*. In the eighteenth century, scientists were uncertain as to just how the male semen managed to fertilize the egg. Two possibilities were recognized:

Hypothesis 1: The semen of the male must make actual contact with the egg before fertilization and embryological development could begin.

Hypothesis 2: It is only necessary that a gas or vapor, arising from the semen by evaporation, make contact with the egg.

From their knowledge of the female reproductive system, as determined primarily through postmortem anatomical dissections, physicians understood that the semen must be deposited a considerable distance from the female ovaries, where the eggs are produced. Because the role played by the sperm cells was not understood, the fact that they were capable of swimming toward the egg was not taken into account. Thus, it seemed reasonable to hypothesize that only a vapor arising from the semen could possibly reach the egg and fertilize it.

In 1785, Italian monk and experimental scientist Lazaro Spallanzani (1729–1799) put the vapor hypothesis to an experimental test, using the toad as his experimental animal. In Table 2.1, Spallanzani's own words from his report on his work provide an excellent example of the underlying logical structure of good scientific procedure.

Spallanzani later performed other experiments that further contradicted hypothesis II, the vapor hypothesis. For example, he discovered that if he filtered the semen through cotton, it lost much of its fertilization powers and that the finer the filter the more these powers were diminished. He also found that several pieces of blotting paper completely removed the semen's ability to fertilize. However, when the portion left on the paper was put into water, it *did* successfully fertilize eggs. Despite the obviousness (to us) of the role played by the sperm in fertilization—a role to which these experiments certainly point—Spallanzani previously had decided that semen without sperm *was* capable of fertilization. He was unable to shake this belief, even in the light of his own experimental results. If nothing else, this demonstrates nicely that scientists are just as prone to overlook the obvious as anyone else and often may refuse to give up preconceived notions despite clear evidence to the contrary. It was not until the nineteenth century that the distinct role of sperm in fertilization was established.

The Logic of Science: Hypotheses as Explanations

Although generalizing hypotheses are important in science, hypotheses that actually *explain* a phenomenon are preferable. In the case of the sour green apples, for example, a hypothesis might be developed to explain why they are sour. One such explanation might be that green apples contain high concentrations of acid. This explanation is readily testable by carrying out chemical analyses and comparing the amount of acid found in sour green apples with that in sweet apples. It is only those hypotheses that actually explain observed natural phenomena that may rise to the level of being considered theories.

Teleological Versus Causal Hypotheses

Throughout the history of biology, two types of hypotheses have been put forward. Teleological (Greek for *telos*, end; meaning goal-oriented) hypotheses suggest that certain events or processes occur for some purpose or are

Spallanzani's experiments on the vapor thesis of fertilization, 1785*

Spallanzani's text	Analysis
Is fertilization affected by the spermatic vapor? It has been disputed for a long time and it is still being argued whether the visible and coarser parts of the semen serve in the fecundation [early development] of man and animals, or whether a very subtle part, a vapor which emanates therefrom and which is called the aura spermatica, suffices for this function.	*Here the problem is defined: Does the semen itself cause the egg to develop? Or is it merely the vapor arising from the semen that does so?*
It cannot be denied that doctors and physiologists defend this last view, and are persuaded in this more by an apparent necessity than by reason or experiments.	*Here, Spellanzani points out the lack of experimental evidence to support the vapor hypothesis.*
Despite these reasons, many other authors hold the contrary opinion and believe that fertilization is accomplished by means of the material part of the semen.	*In the full text of his report, he cites some of the anatomical observations noted in the introductory part of this section.*
These reasons advanced for and against do not seem to me to resolve the question; for it has not been demonstrated that the spermatic vapor itself arrives at the ovaries, just as it is not clear whether the material of the semen that arrives at the ovaries, and not the vaporous part of the semen, is responsible for fertilization.	*He next states the alternative hypothesis: that the semen must actually make contact with the egg for fertilization to occur. Two alternative hypotheses now can be tested. The statement "it has not been demonstrated that" again shows Spallanzani's recognition of the lack of empirical evidence to support or refute either hypothesis.*
Therefore, in order to decide the question, it is important to employ a convenient means to separate the vapor from the body of the semen and to do this in a way that the embryos are more or less enveloped by the vapor;...	*An experimental design is suggested. Some sort of apparatus must be constructed to properly test the two alternative hypotheses.*
...for **if** they are born, [**then**] this would be evidence that the seminal vapor has been able to fertilize them; or [**if**] on the other hand, they might not be born, **then** it will be equally sure that the spermatic vapor alone is insufficient and the additional action of the material part of the semen is necessary.	*Note the occurrence here of the if...then format, as Spallanzani identifies the deductive logic of his experimentation. He had shown earlier that the semen could be diluted several times, yet still remain capable of fertilization. In terms of what is known today, this is not surprising. However, Spallanzani interpreted these results as support for the vapor hypothesis, because he considered vapor to be merely diluted semen.*

Spallanzani's text	Analysis
In order to bathe the tadpoles [eggs] thoroughly with this spermatic vapor, I put into a watch glass a little less than 11 grains of seminal liquid from several toads. Into a similar glass, but a little smaller, I placed 26 tadpoles [eggs] which, because of the viscosity of the jelly [the eggs of amphibians such as frogs and toads are surrounded by a jelly coat] were tightly attached to the concave part of the glass. I placed the second glass on the first, and they remained united thus during five hours in my room where the temperature was 18° [Celsius]. The drop of seminal fluid was placed precisely under the eggs, which must have been completely bathed by the spermatic vapor that arose; the more so since the distance between the eggs and the liquid was not more than 1 ligne [2.25 mm]. I examined the eggs after five hours and found them covered with a humid mist, which wet the finger with which one touched them; this was however only [the] portion of the semen which had evaporated and diminished by a grain and a half. The eggs had therefore been bathed by a grain and a half of spermatic vapor; for it could not have escaped outside of the watch crystals since they fitted together very closely...But in spite of this, the eggs, subsequently placed in water, perished.	*Like many in his day, Spallanzani believed in the preformation theory, proposing that the animal egg contained a miniature of the adult form that needed only fertilization to grow to full size; hence his reference to the unfertilized eggs as tadpoles. One experiment, however, seems to have convinced him otherwise. Spallanzani here describes his experimental set-up (Fig. 2.5). The fact that development did not occur meant that the prediction that necessarily followed from the acceptance of the hypothesis being tested did not come true, and thus the vapor hypothesis must be false.*
Although the experiment overthrows the spermatic vapor theory...it was nonetheless unique and I wished to repeat it.	*Spallanzani recognizes the need for additional experimental evidence demonstrating that the vapor hypothesis is, indeed correct. His results in a second series of experiments were the same.*
Having previously used spermatic vapor produced in closed vessels, I wished to see what would happen in open vessels in order to eliminate a doubt produced by the idea that the circulation of the air was necessary for fertilization...	*He recognizes a variable factor that might influence the results, and modifies the experiment to eliminate this variable. If air played a role in fertilization, then the eggs should develop if air was allowed to circulate.*

Spallanzani's text	Analysis
…but fertilization did not succeed any better than in the preceding experiments.	*Again, negative results. The prediction is shown to be false.*
The last experiment of this type was to collect several grains of spermatic vapor and to immerse a dozen eggs in it for several minutes; I touched another dozen eggs with the small remnant of semen which remained after evaporation, and which did not weigh more than half a grain; eleven of these tadpoles [eggs] hatched successfully although none of the twelve that had been plunged into the spermatic vapor survived.	*He performs yet another variation of the original experiment. This experiment yielded additional evidence against the vapor hypothesis: Even immersion in the condensed spermatic vapor did not result in fertilization! If the vapor hypothesis were valid, different results certainly would have been predicted.*
The conjunction of these facts evidently proves [supports the view] that fertilization in the terrestrial toad is not produced by the spermatic vapor but rather by the material part of the semen.	*Expressed in deductive format, Spallanzani's results shows the vapor hypothesis to be false. Despite his use of the word "proves," these results did not mean that the alternative hypothesis was correct, but only provides support for it.*
As might be supposed, I did not do these experiments only on this toad, but I have repeated them in the manner described on the terrestrial toad with red eyes and dorsal tubercles, and also on the aquatic frog, and I have also had the same results. I can even add that although I have only performed a few of these experiments on the tree frog, I have noticed that they agree very well with all the others.	*Note also that Spallanzani is careful not to generalize beyond the species of animal used in his experiments.*
Shall we, however, say that this is the universal process of nature for all animals and for man?	*Spallanzani now wishes to extend his results to other organisms and so performs other experiments using different species. In other words, can the generalization be extended to other organisms not yet tested in these experiments?*
The small number of facts which we have does not allow us, in good logic, to draw such a conclusion. One can at the most think that this is most probably so…	*Spallanzani is properly cautious in considering an extension of his generalization about the necessity of contact with the semen (rather than its vapor), beyond the small group of species on which he*

Spallanzani's text	Analysis
	actually carried out his experiments. He shows his awareness of scientific logic with this statement.
...more especially as there is not a single fact to the contrary...and the question of the influence of the spermatic vapor in fertilization is at least definitely decided in the negative for several species of animals and with great probability for the others.	*No false predictions are obtained in the experimental testing of this hypothesis (that is, hypothesis 1), but the conclusion still could be expressed only in terms of probability. Note Spallanzani's awareness that his negative results give positive disproof of the vapor hypothesis, yet provide only probable verification, rather than absolute demonstration that this is the case with other species.*

*Excerpts from Spallanzani's accounts of his experiment are reprinted from M. L. Gabriel and S. Fogel, eds. *Great Experiments in Biology.* Englewood Cliffs, NJ: Prentice-Hall; 1955. All brackets in Spallanzani's accounts of his experiments have been inserted by authors for clarity.

directed toward some end. In contrast, causal hypotheses focus specifically on the direct factors that lead from event A to event B.

Consider the following example. More than 40 years ago, Harvard University biologist Ernst Mayr observed that a warbler living all summer in a tree next to his house in New Hampshire began its southern migration on August 25. This single observation raised a question in his mind. *Why* did the warbler begin its migration on that date? A teleological answer to this question might be: "Because the warbler wanted to move to a warmer climate." Teleological answers imply conscious, goal-oriented behavior. Although it may be appropriate to ask teleological questions about human

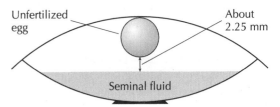

Figure 2.5 Experimental set-up similar to the one used by Spallanzani to answer the question "Is fertilization effected by the spermatic fluid?" Vapor rising from the seminal fluid freely bathed the egg, but no contact between egg and fluid occurred. The egg did not become fertilized.

activities, it is futile to ask such questions about other organisms. To ask "For what purpose did the warbler begin to migrate?" implies that the warbler made the same kind of conscious, goal-oriented choice that a person might make in deciding to go shopping. There is no way to test teleological hypotheses scientifically. One obviously cannot ask a warbler to tell us its reasons for leaving New Hampshire in late August! This is why biologists seek nonteleological, or causal, hypotheses that focus on more specific, testable reasons that a bird might begin its migration at a certain time each year. Mayr's original question can be rephrased in a more precise manner: "What factor or factors caused the warbler to begin its migration on August 25?" It is possible to answer this question without making ungrounded assumptions about a conscious purpose on the part of the warbler in starting its migration.

Before we try to answer Mayr's question, we need further clarification. His original question was phrased in the singular: "What caused *the warbler* to begin its migration on August 25?" In science, such questions are more likely to be posed in a general form: "What causes *warblers* to begin their migration *around* August 25?" No two warblers are the same, and no two August 25ths have absolutely identical conditions. Although biologists investigating such problems may have to deal with individual cases, it is important to frame the question in as broad a manner as possible. In general, scientists are more interested in explaining the principles underlying collections of events than in accounting for individual events, such as the behavior of a single warbler. The greater the number of warblers studied, the greater the likelihood of validity for any hypotheses generated about the migratory behavior of warblers in general.

Types of Causal Explanations

In response to the general question on the causes of warbler migration, at least three different kinds of causal explanations are possible (Fig. 2.6):

An internal hypothesis. Warblers begin migrating toward the end of August because a physiological mechanism (for example, a hormonal change) is activated, leading to migratory flight behavior. This explanation focuses on a physiological mechanism within the organism that may trigger migratory behavior.

An external hypothesis. One or more specific environmental factors, such as the short day length associated with fall or a decline in the insect population that constitutes the warbler's food supply, may activate the

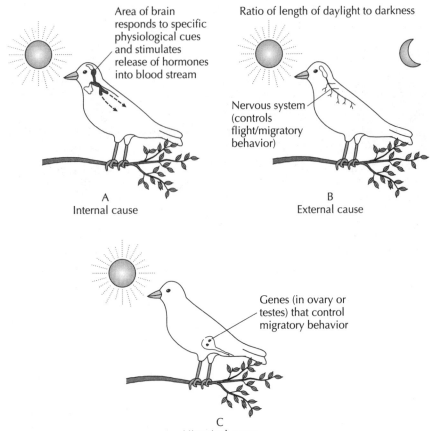

Figure 2.6 Three possible causal explanations for why a bird might begin its migratory flight on August 25.

physiological trigger for migration described in the first hypothesis. The external hypothesis emphasizes factors outside the organism that may trigger migratory behavior.

A historical hypothesis. Warblers begin moving south because, through the course of evolution, they have acquired a genetic constitution that programs them to respond to certain environmental changes associated with the end of summer. This historical hypothesis relates warbler migration to an adaptive response to environmental changes developed through evolutionary processes over long periods of time.

These three types of explanations are not mutually exclusive, and the most complete explanation may well involve aspects of all of them.

Cause and Effect

As the preceding discussion suggests, modern science is built on belief in cause and effect: For every observed effect, there is some cause or set of causes. Yet, causal hypotheses have distinct limitations. One lies in our ability to test them. For example, the hypothesis proposing that warbler migration is the result of a delicate change in hormones is reasonable, but, if a technique for measuring small hormonal changes within the organism is not available, the hypothesis remains untestable and therefore is of limited value. Further, many cause-and-effect relationships may be more apparent than real. Suppose, for example, that a cold air mass from Canada arrived in New Hampshire on August 25, just when the day length was appropriate to trigger the warbler's migratory response. It might *appear*, therefore, that the primary cause for onset of migration was the drop in temperature, an example of a spurious cause. Such spurious relationships are common in nature, for many events occur simultaneously and often may seem to be causally related.

Bias in Science

Science is not an abstract process isolated within the ivory towers of colleges, universities, and research institutes. Rather, science is always situated in a social context in which economic, political, and philosophical values influence everything from the precise nature of the research undertaken to the actual kinds of hypotheses considered acceptable. Like all people, scientists grow up within specific cultures and learn to accept certain values related to their time and place in history. These values often may represent general biases that people in a given society share. Individuals also have their own personal biases. For example, some biologists are biased against explanations that cannot be expressed in mathematical or molecular terms, whereas others believe that such explanations lose sight of the fact that the whole organism is the most significant unit of biological function. Religious and political biases also may play roles in influencing individual scientists. Despite stereotypes to the contrary, individual scientists are no less prone to personal biases than anyone else.

Two kinds of bias may appear in scientific work. One is conscious bias, a deliberate manipulation or alteration of data to support a preconceived idea. Conscious or intentional bias in science is, in reality, simple dishonesty. Because science is based on repeatability, such dishonesty is relatively rare; fraudulent work eventually will be uncovered by other investigators. Far more common is a bias of which the individual may be totally unaware Because science is a human activity, unconscious bias is almost inevitably present to some extent in research. To present a more realistic view of the scientific process, it is important to examine the way in which biases of various kinds function in science.

First, let us clarify what we mean by "bias." In the English language, the term has a negative connotation, implying an undesirable component of the thinking process. When we say that a scientist may be "biased," we also may be saying that he or she may have a particular point of view that influences the selection and formulation of hypotheses. Sometimes these points of view act as blinders, preventing scientists from seeing the value of new ideas. At other times, however, points of view also may act as catalysts, providing new insights or ways of looking at problems. To understand both the positive as well as the negative roles that bias may play in science, let us examine briefly one example involving conscious bias and another involving unconscious bias.

Conscious bias. The theory of inheritance of acquired characteristics holds that traits acquired by an organism in its own lifetime (for example, a large body musculature acquired by exercise) may be passed on to offspring. By the twentieth century, virtually all biologists had rejected this concept, which dated from the early nineteenth century. A few researchers, however, persisted in trying to show that acquired characteristics could be inherited.

One of these researchers was Austrian biologist Paul Kammerer (1880–1936). In the years immediately after World War I, Kammerer studied inheritance in the midwife toad (*Alytes obstetricans*). Most toads and frogs mate in the water. To grasp onto the female during mating, males develop rough "nuptial pads" on their palms and fingers during the mating season, enabling them to hold onto the slippery female. However, because midwife toads mate on land and the female's skin is rough and dry, no selective advantage accrues to males possessing nuptial pads, so the pads are absent in *Alytes*. Kammerer, who was very talented at raising and handling amphibians, was able to induce midwife toads to mate in water. After only a few generations, he claimed that the males showed nuptial pads and, far more significant, transmitted this trait to their male offspring.

Kammerer's results received criticism from many quarters. For one thing, for all his skillful experimental abilities, Kammerer was not a good photographer. Many of the published photos of his specimen were difficult to see or appeared to be retouched. For another, visitors to Kammerer's lab in Vienna were never able to see a live specimen with nuptial pads. The final blow to his hypothesis came when several biologists, including American G.K. Nobel, examined the one surviving preserved specimen. Nobel observed that the so-called nuptial pad was a blotch on the skin induced by what appeared to be the injection of India ink under the surface.

Some evidence suggested that Kammerer himself did not fake the data and that the injections were made by one of his assistants. Even so, had he not been such a zealous advocate for the idea of inheritance of acquired characteristics, Kammerer might have examined his own results more carefully and avoided public discredit by the episode. Whether this revelation of fraud was responsible for his suicide some months later has never been established. He was also involved at the time in an unhappy love affair, but it seems likely that the accusations of fraud must have played some role in his death.

Unconscious bias. Unconscious bias is less straightforward than conscious bias and is far more common in scientific work. We will discuss just one example.

For thousands of years, humans had been aware that meat left out in the open would soon contain maggots, the larval stage of flies. The hypothesis put forward in the seventeenth century to explain this observation was that the maggots were spontaneously generated from the organic matter of the meat in contact with air. This hypothesis led to the prediction that any meat placed out in the open would soon show the presence of maggots, a prediction confirmed by experience.

An alternative hypothesis, however, had also been proposed: that maggots developed from eggs laid by adult flies on the meat. This was not a trivial or purely academic issue, because infestation of meat by maggots in open-air markets was a major economic and health problem in the seventeenth century. If maggots really came from flies, then a method of protecting meat suggested itself immediately. In light of this alternative explanation, the Italian naturalist Francesco Redi (1636–1697) realized that he could test the spontaneous generation hypothesis with a simple experiment:

Hypothesis: *If* . . . spontaneous generation is responsible for the appearance of maggots in meat exposed to air and,

if . . . meat is exposed to air in a jar covered by gauze to exclude adult flies,

Prediction: *then* . . . maggots should still develop in the meat.

Redi set up the experiment shown in Figure 2.7. He used two jars, into each of which he placed the same amount and type of meat of the same age and obtained from the same butcher. One jar was left open (Fig. 2.7A), and the other was covered with gauze (Fig. 2.7B). The first jar served as a control in this experiment, and the second served as the experimental jar. The experimental element is that which has been modified (in this case covered with gauze) to test a specific hypothesis, and the control remains unmodified to serve as a comparison. For example, if the day of the experiment was quite cold and there happened to be no flies around, then the meat in the control jar should not develop maggots either. In setting up controls for experiments, it is important that all factors except the one being tested are kept the same. Thus Redi used the same kind of meat of the same age from the same butcher (note that if the meat already contained fly eggs when it was bought, the experimental results would obviously be invalidated). The two jars would have to be placed in the same part of the room, kept at the same temperature, etc. As the experiment proceeded, Redi observed flies hovering around the tops of both jars. Because jar A was uncovered, the flies could enter and come in contact with the meat, but, because of the gauze, they could not get inside jar B. The meat in jar B eventually spoiled through bacterial decay, but no maggots

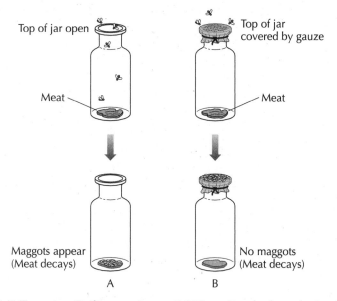

Figure 2.7 Francesco Redi's experiment (1685) testing the hypothesis of spontaneous generation of maggots from meat exposed to air.

appeared. The meat in jar A developed maggots. Redi thus was able to show that the hypothesis of spontaneous generation of maggots led to a false prediction and, therefore, could be rejected.

Or could it? History shows that the outcome was not so simple. Proponents of the spontaneous generation hypothesis had a comeback. They argued that the gauze changed the *quality* of the air getting to the meat, thereby preventing the normal process of maggot generation from occurring. Additional experiments and observations—for example, observing flies actually laying eggs on the meat and microscopic observation of eggs developing into maggots—were necessary before this modified version of the hypothesis of spontaneous generation also could be rejected.

Despite Redi's experiments, biologists in the 1860s were again debating the old issue of spontaneous generation, this time in France. In this version, however, the debate occurred in a different context. Attention now was focused on the spontaneous generation of bacteria that routinely appeared in milk, beer, and wine, causing them to sour. The germ theory of disease, championed by French microbiologist and chemist Louis Pasteur (1833–1895), was just gaining ground at this time. This theory proposed that the souring was caused by the presence of bacteria. An important component of Pasteur's germ theory was his opposition to spontaneous generation of any sort. All bacteria, he claimed, came from the reproduction of previously existing bacteria and none were spontaneously generated from nonliving organic matter.

An opposing hypothesis, invoking spontaneous generation, was put forth by another French biologist, Felix A. Pouchet (1800–1873). He argued that, although bacteria were certainly able to reproduce themselves, they also could be formed spontaneously from the right combination of organic materials. Pouchet performed a simple set of experiments that appeared to support his spontaneous generation hypothesis. He found that if a series of flasks of hay infusion were heated to about 100° C, so as to sterilize their contents, and then were sealed and left at room temperature, after a short time they were seen under the microscope to be teeming with bacteria.

In response to Pouchet's claims and in defense of his own view, Pasteur countered that bacteria existed everywhere—on hands, in food, wine, beer, and milk—and could be carried by the air from place to place. He argued that Pouchet either had not fully sterilized the original broth or had not plugged up the flasks quickly or carefully enough, thus allowing bacterial contamination of the liquid. This contention led Pouchet to repeat his own experiments several times, always with the same results.

The two scientists proceeded to exchange comments and letters in scientific publications. So intense was the debate that, in 1861, the French Academy of Sciences arranged a series of public presentations and a contest on the topic, with a cash prize to be awarded by a jury of scientists to the best presentation. Pasteur and Pouchet were the main contenders. For his part, Pasteur first demonstrated that boiling beef broth in a flask and then immediately sealing it (by melting the glass at the top so that the contents were not allowed contact with air) prevented the growth of bacteria and hence decay of the broth. (Bringing organic material close to the boiling point is the basis for "pasteurization," now used routinely to prevent the souring of milk, wine, beer, and other foods.) However, Pouchet countered with a perfectly reasonable argument: Boiling might have changed the chemical composition of the organic material in the broth, as well as the air inside the flask, rendering both unsuitable for the spontaneous generation of microbes.

In response to this claim, Pasteur performed another simple but elegant experiment. He boiled beef broth in a specially designed long-neck flask (Fig. 2.8A) that allowed air to diffuse back and forth between the broth and the outside. The lower portion of the neck of the flask served as a trap for the heavier dust particles and bacteria carried in the air. Using this apparatus, Pasteur reasoned, would allow air to come in contact with the broth, but no airborne bacteria would make it beyond the trap. If spontaneous generation could occur, then it ought to do so under these circumstances. The results of Pasteur's experiment were quite dramatic. Even after several months, there was no decay in the flask. Moreover, he made another bold prediction: *If* bacteria were airborne and getting caught in the "trap," and *if* he tilted the flask so that some of the broth got into the trap and was then returned to the main receptacle (Fig. 2.8B), *then* the broth in the receptacle should show bacterial growth. When he carried out this experiment, bacteria appeared in the broth in just a few days, as his hypothesis predicted. To Pasteur, this was strong support for his hypothesis and a clear rejection of Pouchet's. The French Academy of Sciences agreed and awarded the prize and membership in the influential academy to Pasteur.

In the course of the debate, Pasteur *seems* to have shown by the sheer force of logic and ingenious experimental design that the theory of spontaneous generation of bacteria could be rejected. Indeed, this is the way the episode is presented in most biographical and textbook accounts. Pasteur himself promoted this interpretation, and his work on spontaneous generation often has been used to illustrate the ideal of pure, unbiased science at work.

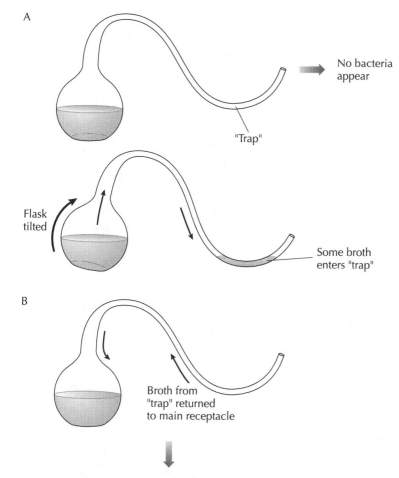

A

No bacteria
appear

"Trap"

Flask
tilted

Some broth
enters "trap"

B

Broth from
"trap" returned
to main receptacle

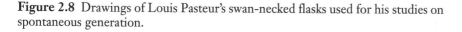

Bacteria appear in broth in 2-3 days

Figure 2.8 Drawings of Louis Pasteur's swan-necked flasks used for his studies on spontaneous generation.

The story is not quite as simple as it might appear at first, however. It is certainly the case that Pasteur's experimental work was exemplary. Yet, other factors also appear to have been at work in motivating his opposition to spontaneous generation. In a detailed study of Pasteur's published and unpublished writings, historians of science Gerald Geison and John Farley have suggested that his position on spontaneous generation was very much influenced by his political and religious views. After the radical revolutions of 1848 that spread throughout Europe, the decade of the 1860s in France was one of growing political conservatism. Pasteur himself was especially conservative in political and religious outlook. In the 1850s he had become an enthusiastic supporter of Emperor Napoleon III (nephew of Napoleon I) and the restoration of the French Empire, which stood for law and order and for the suppression of "radical" ideas such as the separation of church and state. Pasteur enjoyed the French government's financial support and, on several occasions, was an invited guest of the Emperor at one of his country estates. Pasteur was an aspiring member of the French Academy of Sciences, then composed of the most pro-government, scientific elite of France. Pouchet was also a member of the Academy but worked in the provincial town of Rouen and thus was not among the inner circle that dominated the Academy in Paris. At one point, Pasteur even ran for the French Assembly (analogous to the United States Congress) as a member of the conservative party.

Pasteur was a devout Roman Catholic. As early as 1850, he had enthusiastically supported Emperor Napoleon III's use of French military power to enable Pope Pius IX to return to the Vatican. Pius IX was strongly interested in restoring the Church to its former place of political prominence in France. He became noted for his condemnation of all tendencies toward what he termed "religious tolerance" and "modernism" and for convening the Vatican I Council meetings in Rome in the 1860s, at which he proclaimed the dogma of papal infallibility (1870). By the early 1860s, both the Church and the French government had formed powerful allegiances to combat any tendencies—political, religious, or intellectual—that appeared to challenge orthodox views.

Geison and Farley suggest that Pasteur viewed the idea of spontaneous generation as a serious challenge to the established religious views of "special creation" then officially supported by the Catholic Church and French state. Charles Darwin's *Origin of Species* had been published in 1859, just two years before the Pasteur–Pouchet debates began. In that same year, Pouchet had published a major work advocating the possibility of spontaneous generation and presenting some of his experiments. Darwin's book had raised the inevitable question of how the first forms of life had originated on earth, and the theory of spontaneous generation of simple forms like bacteria *seemed* to provide an answer. Thus Pouchet's ideas were viewed by many of his contemporaries, including Pasteur, as advancing the cause of materialism and atheism and, by extension, supporting attacks on the fundamental tenets of established religion.

In a lecture in 1864 at the Sorbonne, perhaps the most famous of the French universities, Pasteur himself made clear his view of the relationship between the theory of spontaneous generation and liberal, "atheistic" ideas. The great question of the day, Pasteur began, was the permanence of species in contrast to the Darwinian idea of their slow transformation. "What a triumph it would be for materialism," Pasteur told his audience, "if it [the theory of evolution by natural selection] could claim that it rests on the [scientifically] established fact of matter organizing itself, taking on a life of its own....To what good, then, would be the idea of a Creator, God?" Although he went on to tell his audience that questions of science cannot be decided by religious doctrine, it seems quite likely that Pasteur's deep-rooted political and religious convictions played an important role in determining which side of the debate he supported. This interpretation is further strengthened by the fact that Pasteur did not bother to repeat Pouchet's most controversial experiments, asserting without providing experimental evidence that the apparent spontaneous generation of bacteria observed by Pouchet must have been the result of contamination of the broth.

The Pasteur–Pouchet case shows that bias may have both a positive and a negative influence on scientific work. On the positive side, Pasteur's opposition to spontaneous generation led him to champion an opposing view, the germ theory of disease, which emphasized that disease may be transmitted by bacteria through personal contact between people or through the air. The germ theory was to have an enormously beneficial impact on medicine and public health in the ensuing decades. Yet Pouchet's view also had its positive side. By promoting the idea that the origin of life could be studied by chemical and physical means, he pioneered a line of inquiry that has become increasingly fruitful and important in biological research. Conversely, Pasteur's opposition to spontaneous generation prevented him from evaluating Pouchet's own evidence more carefully and from seeing that the question of the origin of life could be approached from a purely chemical and physical point of view. Ironically, although he privately entertained the *possibility* of a chemical explanation for the origin of life, Pasteur did not pursue such possibilities extensively, nor did he repeat them publicly. Similarly, Pouchet's advocacy of spontaneous generation led him to deemphasize the importance of transmission of bacterial infection and thus the significance of the germ theory of disease.

Perhaps the greatest irony in this story is that *both* Pouchet and Pasteur were right. You may have noted in the description of their experiments that the two men were using different sources of organic broth. Pasteur's was beef broth, whereas Pouchet's was a hay infusion, a liquid prepared by soaking hay in water. Because he did not repeat Pouchet's experiment, Pasteur did not discover that the natural bacteria found in hay infusions include some species that can form spores, which enable them to survive severe conditions like drought, cold, or heat. The spores present in Pouchet's preparation were heat resistant enough to survive the short boiling time to which he subjected them, thereby emerging from their dormant stage and beginning to reproduce once the flask cooled down. Several decades would pass before the existence of heat-resistant spores was recognized by microbiologists.

The Concept of Paradigms

In June 2000, the National Aeronautics and Space Administration (NASA) announced that new photographs, taken by an orbiting satellite, provided evidence of subsurface water on Mars. This water is thought to break through occasionally and erode the Martian surface, creating meandering, river-like channels revealed by satellite and other photographs.

Because the presence of water on Mars suggests strongly the possibility of past or present simple forms of life, the NASA announcement received wide coverage in the popular press. However, as one astronomer noted, the discovery did not mark a major "paradigm shift" in his field. In essence, what the astronomer was conveying by this comment was that the announcement was "no big deal," and that, although the new pictures may have provided greater detail than before, the presence of channels on Mars had been known for a number of years. Thus, the NASA announcement was not that new to anyone familiar with the field of Martian geography.

The expression "paradigm shift" is one that has received increasingly wide use over the past few decades. In his 1962 book, *The Structure of Scientific Revolutions*, historian Thomas Kuhn (1922–1996) introduced the term "paradigm" (from Greek, *paradigma*, pattern) to refer to a broad collection of ideas, assumptions, and methodologies that guide research in any field of science. Kuhn recognized that, once established, paradigms often resist change, even in the light of considerable contradicting empirical evidence. When the evidence against an established paradigm reaches a certain crucial level, a "scientific revolution" or what Kuhn called a paradigm shift occurs, and a new way of viewing the world or a specific set of problems emerges. Kuhn noted that paradigm shifts may be on a large scale (for example, from the geocentric to the heliocentric view of the universe) or more restricted (for example, from viewing the blood as ebbing and flowing to seeing it as circulating in a one-way path throughout the body). According to Kuhn, all paradigm shifts, large or small, share certain characteristics and reveal a number of important features about how science is practiced. Before listing some defining characteristics of paradigms and paradigm shifts, it is useful to examine two historical examples.

Darwin and the Theory of Evolution by Natural Selection

One of the most profound shifts in our way of thinking about the living world came with the publication in 1859 of Charles Darwin's *On the Origin of Species*, in which he put forward his theory of the transformation of animal and plant species over time by the mechanism of natural selection.

Since the earliest written records, human beings have wondered how the myriad types of animals and plants found on earth—what we today refer to as biodiversity—could have arisen. In western culture, from the ancient world through the nineteenth century, the traditional explanation for biodiversity has been the doctrine (paradigm) of special creation. The various organism types, called species (borrowing Plato's terminology for ideal, fixed categories), were thought to be each a separate entity, created by God in their present forms, stable and unchanging (immutable) over time. Although species were recognized as separate entities, it was also apparent that they could be grouped into similar types (for example, as members of the dog or cat families, or of oak and pine tree families) that shared many characteristics in common. Such groupings were interpreted under this older paradigm as representing God's plan for "Nature." In addition to being part of religious dogma, this older paradigm of biodiversity also formed the basis for the work of important naturalists such as Karl von Linné (1707–1778). Known as Carolus Linnaeus, or simply Linnaeus (the Latinized name under which he wrote), he developed a widely used classification system that grouped species by shared traits.

By 1800, however, the older paradigm (which we may call the Linnaean paradigm, because that is the explicit form in which most naturalists would have encountered it) faced some unexplained observations. Kuhn, in his work on paradigms, referred to such unexplained observations as "anomalies." For one thing, geologists had unearthed multitudes of fossils. Some of these fossil forms were strikingly similar to forms still living today, whereas others such as dinosaurs, and feathered reptiles seemed bizarre and had no living counterparts. Why should so many forms have perished? Why were large groups of organisms, such as all vertebrates or all flowering plants, built on the same basic plans? Why were there certain patterns of geographic distributions of organisms rather than a random distribution?

From 1831 to 1836, Darwin traveled around the world as an unpaid naturalist on board a British exploratory vessel, H.M.S. *Beagle*. He saw an enormous variety of life in an equally enormous range of habitats. From this experience and from copious reading, Darwin gradually came to accept the idea that species were neither fixed nor immutable and that the species we see on earth today must have descended with modifications from species that existed in the past. For example, species that share many characteristics in common, such as the orchids, must all have descended from a common ancestor. Darwin thus embraced the theory of transmutation of species (today known as evolution), which a few other naturalists of his time had already put forth in one form or another, although not very successfully.

Darwin went on to propose a second theory, that of natural selection, as the mechanism for the way in which evolution might occur. The theory of natural selection was based upon several observations: (1) More organisms are born than can survive, resulting in competition for scarce resources. (2) All organisms vary from one another. Those organisms with favorable variations will survive or be more vigorous and thus more fit for the environment in which they live. (3) The more offspring that reach reproductive maturity, the more fit that individual is said to be. (4) Many of these offspring will carry the favorable variations they inherited from their parents and, in turn, will leave more surviving offspring than other members of the population. In this way, the new and favorable variation will spread through the population. (5) Because variations occur more or less randomly, if two portions of a single population are somehow isolated from each other for long periods of time, different variations will accumulate in each population, producing divergence in their characteristics. The result will be that two species eventually will have been derived from one common ancestral form, an evolutionary process known as speciation. Continuous speciation has led to the wide variety of organisms that have populated the earth in the past and today.

Darwin's paradigm challenged every aspect of the older Linnaean paradigm. Species were no longer viewed as fixed and immutable but plastic and ever-changing. They arose not by a supernatural act of creation but by natural processes going on every day. Species were not composed of a single "type," representative of the whole, but by populations that exhibited a range of variability for every trait. Extinction was not a result of God's displeasure with species but of the constant struggle for existence and competition for material resources among various species.

The shift from the Linnaean to the Darwinian paradigm was neither easy nor quick. Many older naturalists simply could not accept the notion of species changing from one form to another; the idea of the fixity of species was too ingrained in their world view. Others gradually accepted evolution (descent with modification from common ancestors) but could not accept the mechanism of natural selection, especially the idea that variations occurred by chance rather than in response to the needs of the organism. Although we often speak of a Darwinian "revolution," most of the initial objections would not be resolved in favor of the Darwinian paradigm until the middle of the twentieth century.

From the start, the Darwinian paradigm encountered enormous opposition from organized religion. As we have seen, even so eminent a scientist as Louis Pasteur found the Darwinian paradigm unacceptable on religious and philosophical grounds. Although many religions found ways to reconcile the Darwinian paradigm with broad theological doctrines, others, especially those who emphasize a literal interpretation of the Bible, continue to object to the evolutionary paradigm even today (see Chapter 5). As Kuhn maintained, paradigm shifts do not come easily.

A Paradigm Shift in Molecular Biology: From the "Central Dogma" to Reverse Transcription

In the 1960s, the understanding that deoxyribonucleic acid (DNA), the molecule that makes up the genes of most organisms on earth, controlled hereditary traits was referred to as the "central dogma" of molecular biology. Despite the negative connotation of the term in scientific circles, it was called dogma because it was supposed to be a universal paradigm. The central dogma stated that DNA exerts its effects by serving as the template for transcribing a second form of nucleic acid, ribonucleic acid (RNA), which, in turn, guides the assembly of a specific protein molecule. The central dogma is often represented as a simple flow diagram:

$$DNA \xrightarrow{\text{Transcription}} mRNA \xrightarrow{\text{Translation}} Protein$$

The first step (first arrow), in which messenger RNA (mRNA) is synthesized from DNA is known as transcription. The second phase (second arrow), in which mRNA guides the production of a specific protein, is known as translation. As it was conceived in the early 1960s, both phases of the process were thought to be unidirectional. The central dogma became the paradigm used to explain how genes function chemically in all organisms.

In 1958, a young graduate student named Howard Temin was working with Rous sarcoma virus (RSV), a virus found in chickens and the first cancer-causing virus to be described. Viruses are very simple structures consisting, in this case, of a protein coat surrounding a nucleic acid core (Fig. 2.9A). It was widely known that RSV belonged to a special group of viruses with hereditary material made up of RNA rather than DNA. All viruses replicate themselves by attaching to a cell surface and injecting their

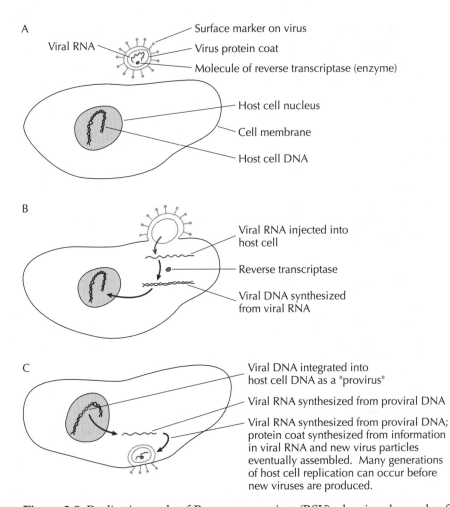

A

Viral RNA
Surface marker on virus
Virus protein coat
Molecule of reverse transcriptase (enzyme)

Host cell nucleus
Cell membrane
Host cell DNA

B

Viral RNA injected into host cell
Reverse transcriptase
Viral DNA synthesized from viral RNA

C

Viral DNA integrated into host cell DNA as a "provirus"
Viral RNA synthesized from proviral DNA
Viral RNA synthesized from proviral DNA; protein coat synthesized from information in viral RNA and new virus particles eventually assembled. Many generations of host cell replication can occur before new viruses are produced.

Figure 2.9 Replication cycle of Rous sarcoma virus (RSV), showing the mode of infection of the host cell by the virus. RSV is an RNA virus, meaning its genetic material is RNA and DNA. By means of surface markers, the virus attaches to the cell membrane and injects its RNA, along with a molecule of the enzyme reverse transcriptase, into the host cell. Reverse transcriptase, as Temin predicted, is able to catalyze the transcription of viral RNA into viral DNA, which is then integrated into the host cell genome as a "provirus." When the host cell's DNA is transcribed to make proteins, it also makes proviral proteins. The presence of these proteins stimulates uncontrolled cell division, producing a cancerous tumor. Moreover, the provirus replicates every time the cell divides, thus increasing dramatically the number of cells with provirus in them. Eventually, enough viral RNA and protein coat material are present so that new viruses can be assembled, breaking out of the cell and infecting other cells nearby.

DNA or RNA into the host cell, where the viral nucleic acid proceeds to commandeer the cell's metabolic machinery to produce more viruses (Fig. 2.9A–C). One of the distinguishing features of RSV is that it does not usually kill the cell it invades but, instead, causes the cell to start dividing uncontrollably, thereby producing a sarcoma, or cancerous tumor. In 1964, while attempting to develop a chemical means of identification (assay) for cells infected with RSV, Temin made several interesting observations: (1) Cells infected with RSV showed recognizable modifications, which were passed on to the progeny cells, even when no further viral replication inside the cell was observed. (2) Substances (such as cytosine rabinoside) known to block DNA synthesis prevented viral infection if applied within 12 hours of the first contact between virus and susceptible cells. (3) Substances (such as actinomycin D) that inhibited the synthesis of RNA from DNA (transcription) allowed infection to occur but blocked viral replication. These data suggested that RSV replication might be a two-step process that somehow involved DNA replication, a rather surprising finding for an RNA-based virus.

To explain these results, Temin proposed what he called the DNA provirus hypothesis, which suggested that when RSV first enters a host cell it uses its RNA as the basis for synthesizing DNA, which now carries viral genetic information. Moreover, Temin argued, the new viral DNA becomes integrated into the DNA of the host cell (in which form it is known as a provirus) and is replicated with the host DNA every time the cell divides. The provirus DNA, when transcribed and translated into protein (by the central dogma pathway), leads to uncontrolled cell division, or cancer. What was so novel—and troubling—about Temin's new paradigm was that it postulated a reversal of the central dogma: In some situations, *RNA could guide the production of DNA*.

The first reaction to Temin's paradigm was almost universal rejection. Not only did it go against the entrenched paradigm of the central dogma, but repeated attempts to detect the presence of proviral DNA in the host cell proved fruitless. Temin continued to search for ways to detect the presence of proviral DNA or other telltale signs of proviral activity. Then, in 1970 Temin and another researcher, David Baltimore, independently reported the discovery of a new enzyme (enzymes are proteins that catalyze biochemical reactions in organisms) in RSV-infected cells. This new enzyme was called reverse transcriptase, because it catalyzes the synthesis of DNA from an RNA precursor. Later, other varieties of reverse transcriptase were found as normal components of animal cells not infected by

RSV. These findings clinched the matter, and, from the mid-1970s on, the provirus and reverse transcription paradigm gained wide acceptance.

Characteristics of Paradigms

It may be useful to ask at this point: Why do we bother to talk about paradigms at all? Why not just refer to Darwin's theory of evolution by natural selection, or Temin's theory of reverse transcription? What has been gained by Kuhn's new terminology and analysis of scientific change?

Paradigms are more comprehensive than theories. A paradigm is a collection of theories but also has embedded in it a variety of assumptions and specific methodologies and is shared by a particular community of investigators. As we have seen, the Darwinian paradigm encompasses not only the theory of evolution but also the theory of natural selection. It also includes theories about the nature of heredity and variation, about adaptation of organisms to particular environments, mate selection, etc. An advantage of seeing a paradigm as consisting of a variety of theories is that it helps us understand how scientific ideas gain acceptance or come to face rejection. For example, Darwin's view that hereditary variations were always very small changes has been challenged by biologists in the past (and even at present), although these critics do not oppose the concept of evolution or natural selection.

Paradigms embody certain methods that are agreed upon as appropriate by the community of investigators in the field. For example, molecular geneticists agreed that, if Temin's paradigm were to be accepted, it was necessary to use biochemical methods to identify an enzyme that would carry out reverse transcription. Agreement on the methods and instruments to be used in a field is one of the main components of a paradigm that knits its adherents together into a social as well as intellectual community. Agreement on methods is also crucial if workers are to evaluate each other's data and discuss interpretations meaningfully.

Paradigms, more than theories in the traditional sense, represent world views, or global ways of seeing nature. A Darwinian paradigm of species transformation is a very different view of the natural world than the old Linnaean view of static, immutable species. The Darwinian world is ever changing, dynamic, never at rest. Change is both expected and celebrated, for it is the means by which organisms survive and adapt to an ever-changing environment. Similarly, the Temin paradigm presents a view of the cell that is more flexible, with a repertory of processes that can meet a larger variety of physiological needs. By contrast, the paradigm represented by the central dogma presents a view of the cell that is more rigid and mechanical. More than just discovering the nature of reverse transcription, Temin's paradigm suggests we should not accept so readily the idea that cells (or any biological system) have only one way of doing things. As one of the characters says in Michael Crichton's book *Jurassic Park*, "Life will find a way."

Kuhn's analysis is particularly helpful in understanding how scientific ideas change. Paradigm shifts are more difficult than merely substituting one theory for another, precisely because paradigm shifts involve a complete change in world view. In Darwin's case, that world view encompassed not only naturalists' conceptions of species as fixed or immutable but the religious doctrine of special creation and the role of God as creator. It is no wonder that the reaction against the Darwinian paradigm was so violent and has been so long lasting. In a smaller way, the paradigm shift from central dogma to reverse transcription initially had its strong opponents, who ridiculed the idea that DNA could be made from an RNA precursor. The shift in world view to reverse transcription cast the fundamental relationship in molecular genetics among DNA (genes), RNA, and proteins in a completely different light. Reorienting that relationship required a shift in world view, at least for those working in molecular genetics.

Old paradigms are replaced with new ones when "dead ends" are reached, for example, when it is simply no longer intellectually satisfying to account for every aspect of species diversity or the fossil record by saying "God made it that way." Old paradigms also tend to be overthrown when they act to restrict rather than expand the scope of the questions being asked. One of the positive outcomes of a paradigm shift is that new areas of research open up, new sets of questions are asked, and research projects are designed to answer them. Resistance to paradigm shifts emanating from within the scientific community merely suggests that scientists, like other specialists, may get so hung up on their own tiny set of problems that they can see the world from only one viewpoint and are blinded to alternatives. In many historical examples, paradigm shifts occur not simply because scientists become intellectually convinced that new evidence is so overwhelmingly in favor of the new paradigm, but because the older scientists retire or die and younger scientists, with fewer ties to the old paradigm, take their place.

Kuhn's analysis also has provided a more realistic understanding of the way science works. An older and more traditional view of science was that it changed by adding more and more information to its basic storehouse of facts. In this view, science is cumulative, with each successive generation getting closer and closer to more accurate views of nature. Also by this view, science was seen as progressive, gaining more and more positively verified and accurate knowledge. Kuhn's view suggests that, although the measurements and data of science in fact may accumulate in some sort of linear way, conceptualizations of science can undergo radical change. Old paradigms are completely discarded and replaced lock, stock, and barrel by new paradigms. A Darwinian view of species is not just a modified Linnaean view. Species and even old data about any given species are now seen in a different light. To Linnaean taxonomists, variations among members of a species were a nuisance that they needed to look beyond to make a proper classification. To Darwinians, variation is the crucial feature by which species evolve to meet new circumstances. Far from being a nuisance, it is a creative force in the development of life.

Modern Science, Materialism, and Idealism

Mechanism and Vitalism

Distinct differences between living and nonliving matter are obvious. The ability to move, ingest materials from the environment, and convert them into more living material is the process we call growth. Growth and, indeed, all of those properties we associate with living organisms are clearly qualitatively different from anything observed in nonliving matter. In the late nineteenth and early twentieth centuries, recognition of this fact led to a resurgence of a much older philosophical debate in the field of biology. This debate was between those who called themselves mechanists and those who called themselves vitalists. Vitalists explained the unique features of living matter by postulating the existence of a "vital force" (*élan vital*). This vital force was assumed to be wholly different from other known physical and chemical forces and ultimately to be unknowable, that is, not subject to physical or chemical analyses. The postulated vital force departed from a cell or organism at death. It should be stressed that vitalists did not deny that chemical analyses of living organisms were valuable but believed that what we call "life" involved something more than could be described by the principles of chemistry and physics.

A classic example of vitalist thinking appears in the American Civil War-era memoirs of Assistant Surgeon Edward Curtis of the U.S. Army Medical Museum in Washington, DC. On April 14, 1865, president Abraham Lincoln was shot in Ford's Theater and died the next morning. An autopsy was performed in the northeast corner guest room of the White House. In the words of Dr. Curtis:

> Silently, in one corner of the room, I prepared the brain for weighing. As
> I looked at the mass of soft gray and white substance that I was carefully

washing, it was impossible to realize that it was that mere clay upon whose workings, but the day before, rested the hopes of the nation. I felt more profoundly impressed than ever with the mystery of that unknown something which may be named "vital spark" as well as anything else, whose absence or presence makes all the immeasurable difference between an inert mass of matter owing obedience to no laws but those governing the physical and chemical forces of the universe and, on the other hand, a living brain by whose silent, subtle machinery a world may be ruled.

This example illustrates clearly a vitalistic belief in some sort of supernatural factor differentiating living from nonliving matter. In brief, the vitalist philosophy may be expressed as one that viewed the whole as being greater than the sum of its parts.

Mechanists, on the other hand, maintained that organisms were simply physical and chemical entities composed of material parts with functions that could be investigated and ultimately explained by the ordinary laws of physics and chemistry. In contrast to vitalists, mechanists viewed organisms as merely complicated machines. Basic to the mechanist view was the idea that organisms are composed of separate parts (molecules, cells, organs) and that we need to only study these parts in isolation to explain the working of the whole. In stark contrast to vitalists, mechanists maintained that the whole *is* equal to the sum of its parts—no more, no less. Thus to learn about how the heart works, for example, a physiologist might remove the heart from an experimental animal, place the heart in a perfusion chamber, in which it could be exposed to fluids with different hormones or chemical transmitters, and measure the effect on the rate of the heartbeat. By such procedures, mechanists thought that all the characteristics of the heart or, indeed, any component of the organism, could be understood.

Experiments by two German biologists played a role in the late nineteenth- and early twentieth-century vitalist–mechanist controversy. Wilhelm Roux (1850–1924) allowed fertilized frog eggs to begin development by dividing into two cells and then killed one of the cells with a red-hot needle. The result was the formation of an incomplete "half" embryo as the living cell continued to divide (Fig. 2.10). Hans Driesch (1865–1941) performed a similar experiment using fertilized sea urchin eggs. However, after the egg divided, instead of killing one of the first two cells, Driesch separated them by shaking the solution vigorously. The result was that both cells formed complete sea urchin larvae (Fig. 2.11). After a decade of using a variety of physical and chemical methods to investigate this problem, Driesch became a champion of vitalism, eventually giving up science completely and becoming a professor of philosophy. To him, the ability of

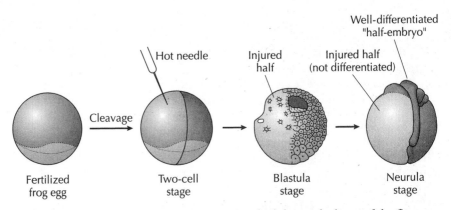

Figure 2.10 Wilhelm Roux's experiment, in which he pricked one of the first two blastomeres of the frog egg with a hot needle, killing it. The result was development of a half-embryo that only reached partial development. These results supported Roux's mechanistic interpretation of development: Each blastomere was already determined at the first cell division to produce half the embryo

the separated blastomeres to adjust to the new conditions imposed on them and still undergo normal development argued strongly in favor of the presence of some inherent nonphysical or chemical force that guided development. He was opposed vigorously in this view by Roux, who developed a very mechanistic research program known as "developmental mechanics." The written arguments between the two scientists became long and occasionally bitter, and, although the two men remained cordial friends until Roux's death, the controversy their work engendered continued well into the twentieth century.

The mechanist–vitalist debate was part of a much large philosophical dispute that has recurred in one form or another from the ancient Greeks to the present day between two broad but mutually exclusive philosophical systems: idealism and materialism. These philosophical terms should be distinguished from their more familiar counterparts, in which idealism refers to an unrealistic and naïve view of the world and materialism refers to an excessive concern with material possessions and wealth. Philosophical idealism and materialism have quite different meanings. Philosophical idealism derives much of its modern content from the writings of Plato and the Platonic tradition in western philosophy. The basic claims are that ideas or nonmaterial causes are the initial and prime movers in the world. Plato saw all material objects, for example, as crude reflections of ideal objects, which existed as a categories in the mind of a Creator. The Linnaean species concept described earlier is an example of idealist thinking,

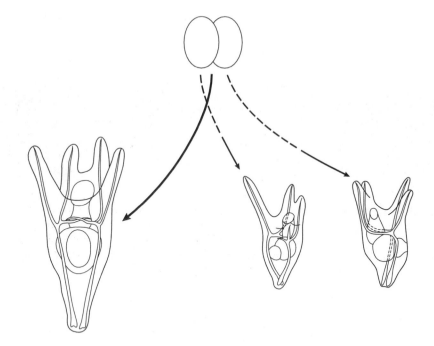

Figure 2.11 Hans Driesch's experiment, separating the first two blastomeres (top) of the sea urchin egg. To the left, a normal sea urchin larva developed from unseparated blastomeres, and, to the right, two slightly smaller but otherwise perfectly formed larvae developed from each separated blastomere. These results contradicted Roux's interpretation by showing that embryos have remarkable abilities to adjust to changed circumstances and are thus not mechanically determined in the way Roux had imagined.

in that each species existed in the mind of the Creator before taking material form on earth. For example, all cats reflect the ideal or essential category of "catness." Real cats are in the world, of course, but the category of "catness" existed before and apart from the appearance of actual cats on the earth's surface and represents the idealized form underlying the group as a whole. Aspects of most contemporary religions are based on idealistic philosophy. Examples include beliefs about a supernatural Creator, the power of prayer, or miracles. Although they may not necessarily express their idealistic views in religious terms, vitalists are clearly idealists in this sense. Idealists do not deny the existence of material reality; they simply relegate it to a secondary role as a causal agent in the real world.

By contrast, philosophical materialism is the view that all processes in the universe are the result of matter in motion. Matter is primary. Everything else, including abstract ideas, is derived from that material reality through our interaction with it. Species, for example, do not exist as an abstract or idealized category apart from the material populations of organisms that form them in nature. We may, of course, create species categories, but they derive from observing actual organisms in nature. Evolution does not occur because of God's plan or abstract "drives toward perfection" but because organisms are competing with one another for scarce material resources. Materialists maintain that nothing is unknowable, although, of course, at any one point in time vast amounts remain unknown. For materialists, the methods of physics and chemistry are the proper tools for understanding the natural world. Modern science since the seventeenth century has rested firmly on a materialist foundation.

In the vitalist–mechanist debates, vitalists clearly were arguing from an idealist position and mechanists from a materialist position. Although both sides made important points, the debate ultimately was counterproductive. Mechanists saw clearly that because, by definition, vital forces were beyond the reach of scientific study, vitalism put limits on scientific research. They therefore rejected it. By the mid-twentieth century, however, many biologists came to the realization that the mechanistic view was too simplistic to account for many functions known to occur in living organisms. Examples include the self-replication of molecules and physiological processes, embryonic development, and a host of others.

Is there an alternative to the mechanistic materialist approach other than vitalism? Biologists and philosophers inspired by biological examples have developed a second form of materialism, holistic materialism, which avoids the pitfalls of both idealism and mechanism. The basic tenets of the holistic approach are:

1. No part of any system exists in isolation. For example, the liver is affected by removing sugar from the blood passing through it, as is the blood by virtue of having that sugar removed. Thus, although both blood and liver may be described partially in isolation, neither can be described fully except in their interactions with each other and all other parts of the body.

2. Unlike machines, living systems are dynamic entities, constantly in a process of change. This change results from the constant interaction of internal and external forces.

3. The internal processes of any living system undergo change as a result of the interaction of opposing forces. All living organisms carry out both anabolic (buildup) and catabolic (breakdown) chemical reactions. For example, the growth and development of a plant

seedling represents a change in which the overall effect of anabolic reactions is greater than that of catabolic reactions. Maturity occurs when the two are balanced, and aging and death result from the dominance of catabolic over anabolic processes. Far from being accidental, this developmental process is programmed into the genetic makeup of the organism. At all stages, the overall process may be studied most fruitfully by investigating the interaction of anabolic and catabolic processes, rather that just anabolic or catabolic processes in isolation.

4. The accumulation of many small quantitative changes may eventually lead to large-scale, qualitative change. For example, the heating of water from 90° C to 91° C represents a quantitative change, because, although it is 1° warmer, the water is still a liquid. However, when the temperature goes from 99° C to 100° C, the water begins to boil and becomes steam. This represents a qualitative change, because water and steam have quite different physical properties. Thus, an accumulation of many quantitative changes has resulted in an overall qualitative change. This is as true of the biological as of the physical world. If a nerve attached to a muscle is stimulated with a low-voltage electric shock, the muscle may not respond. With an increase in voltage, a quantitative change, a qualitative change is eventually achieved in the nerve, an impulse is transmitted, and the muscle contracts.

If organisms are viewed as described in these four tenets, then the vitalists were correct: The whole *is* greater than the sum of the parts. However, an unwarranted pessimism is inherent in the vitalists' position: that because one cannot accurately consider any one part of a living organism in isolation, attempting to completely understand that organism is hopeless. The conflict between mechanism and vitalism was in many ways an artificial one, because both represented limited ways of viewing living systems. Vitalism is limited because it presupposes in living organisms a mystical, unknowable force that, by definition, lies beyond scientific investigation. Mechanistic materialism, on the other hand, in trying to ground biological investigations in knowable but separate physical entities, found it impossible to account for the holistic properties of complex systems like living organisms. The advent of the computer and the mathematical theory of systems analysis have provided important new techniques for studying the complex interactions characterizing living organisms in a rigorous, materialist framework. Biologists increasingly are adopting a more holistic yet still materialist approach that transcends the old mechanist and vitalist limitations.

Conclusion: The Strengths and Limitations of Science

Since its origins in the seventeenth century, western science has proven to be by far the most powerful way of understanding the natural world. As we have seen, one of the greatest strengths of science lies in its emphasis on logical thinking and on formulating testable hypotheses in ways that lead to accurate or refutable predictions. A second strength, growing out of the first, is the insistence that hypotheses and the predictions that follow from them must be testable. A third strength of science is its emphasis on repeatability. A single empirical test, even one repeated by the same experimenter many times, is seldom enough to convince most scientists. Other scientists must be able to replicate the results under the same conditions. Thus, a fourth strength of science lies in its critical, self-policing nature. By checking their own results as well as those of others, any errors are far more likely to be uncovered.

As powerful as it may be as an intellectual tool, science also has distinct limitations. For one, science is limited to dealing with observable phenomena, the empirical data on which hypotheses must ultimately be based. Without empirical data, even the most intriguing hypothesis has little value. Thus, for example, science has absolutely nothing to say one way or the other about the existence of a God or gods or of a human soul. Empirical data on such entities simply are not obtainable. Similarly, science also is limited by the availability of the tools and techniques by which scientific data are gathered. Before the development of the microscope, for example, we knew nothing of the microscopic world. Before the telescope, we could only speculate about the universe beyond what our unaided eyes could see.

Through the self-critical process that characterizes their activity, scientists are forced to constantly modify and many times reject cherished hypotheses. This is *not* because scientists are necessarily more honest and conscientious than persons in other fields but rather because that honesty is reinforced in science by virtue of its system of peer review. All scientists are well aware that others in their field are monitoring their research to make certain their results can be replicated. No scientific idea is likely to remain unchallenged or unchanged. Indeed, scientists probably are wrong more often than right. The nineteenth-century physiologist Johannes Müller (1801–1858) asserted that the velocity of a nerve impulse would never be measured. Six years later, his student, Hermann von Helmholtz measured it in a frog nerve only a few centimeters long. The chemist Ernest Rutherford (1871–1937) stated that the energy in the atomic nucleus would never be tapped; the first atomic bomb exploded just 7 years after his death.

If, as we have seen, scientists can be wrong, so can scientific hypotheses. The fact that science cannot "prove" anything and that its hypotheses and theories are always open to rejection or modification appears to be a limitation. In fact, this limitation is actually the greatest strength of science. According to biologist Garrett Hardin:

> It is a paradox of human existence that intellectual approaches claiming the greatest certainty have produced fewer practical benefits and less secure understanding than has science, which freely admits the inescapable uncertainty of its conclusions.

Hardin is correct: The strength of science does not lie in any claim to infallibility but rather in being an ongoing intellectual process with no pretense of providing final answers or absolute truths. Nor does this strength lie solely in its logical underpinnings, because the conclusion of a perfectly logical argument may well be utter nonsense. The inherent self-criticism of science and its constant search for a better understanding of the natural world through the elimination of false hypotheses are the sources of its immense intellectual power.

Exercises for Critical Thinking

2.1. Distinguish between observation, fact, hypothesis, and conceptualization in science.

2.2. Each statement below can be described as an observation, fact, or conceptualization. Indicate which of these three possibilities best characterizes each statement and explain your choice.

    a. All 100 observers agreed that the sun rises in the east every day.

    b. This green apple is sour.

    c. Planets move from west to east against the background of fixed stars because they are revolving around the sun, just as the Earth does.

    d. The United States fought in Vietnam to preserve democracy from communist aggression.

    e. All green apples are sour.

    f. The report said the witness was lying.

2.3. Devise hypotheses to account for the following observations, and design an experiment or suggest further observations to test your hypotheses.

a. More automobile accidents occur at dusk than at any other time of day.

b. When glass tumblers are washed in hot soapsuds and then immediately transferred face downward onto a cool, flat surface, bubbles at first appear on the outside of the rim, expanding outward. In a few seconds they reverse, go under the rim, and expand inside the glass tumbler.

c. In mice of strain A, cancer develops in every animal living longer than 18 months. Mice of strain B do not develop cancer. If the young of each strain are transferred to mothers of the other strain immediately after birth, cancer does not develop in the switched strain A animals, but it does develop in the switched strain B animals living longer than 18 months.

2.4. A biologist reported the following set of observations:

> While sitting on my porch during the afternoon and early evening, I could not help but notice the chirping of the crickets. I noticed that the rate of their chirping slowly diminished as the sun approached the horizon. I wondered why this should be so. At first I guessed that it was due to the loss of light as the sun disappeared and night came on. Accordingly, I counted the number of chirps given by ten individual crickets in the laboratory as they were exposed to less and less light intensity.

> The following data were obtained:
>
> | | |
> |---|---|
> | At 10 candlepower: | 48 chirps per minute |
> | At 8 candlepower: | 46 chirps per minute |
> | At 6 candlepower: | 49 chirps per minutes |
> | At 4 candlepower: | 55 chirps per minute |
> | At 2 candlepower: | 47 chirps per minute |
> | At 0 candlepower: | 48 chirps per minute |

a. Formulate the hypothesis being tested in a deductive syllogism (*if...then* format).

b. Do the results confirm or reject the second hypothesis? Explain your answer.

The biologist continued:

> I then wondered if temperature might be affecting the number of chirps. I reasoned that, as the sun set, the temperature would drop as night approached. Accordingly, in the laboratory I kept the

crickets exposed to constant light, but exposed them to varying temperatures.

The following data were obtained:

At 34° C:	55 chirps per minute
At 30° C:	48 chirps per minute
At 26° C:	39 chirps per minute
At 22° C:	20 chirps per minute
At 18° C:	8 chirps per minute
At 14° C:	no chirping

   c. In this second experiment, what hypothesis is being proposed and what prediction(s) is (are) made from it?

   d. Do the results confirm or reject the second hypothesis? Explain your answer.

2.5. Indicate for each of the following deductive syllogisms whether they are examples of valid or invalid reasoning, and whether each represents a true conclusion deriving from a true hypothesis(es), a true conclusion deriving from a false hypothesis(es), a false conclusion deriving from a false hypothesis(es), or a false conclusion deriving from a true hypothesis(es)

   a. *If* . . . all residents of the United States are Martians, and
     if . . . all Martians pay taxes,
     then . . . all residents of the United States pay taxes.

   b. *If* . . . all dogs have four legs, and
     if . . . this animal is a dog,
     then . . . this animal has four legs.

   c. *If* . . . all residents of the United States are Martians, and
     if . . . all Martians are green and have tentacles,
     then . . . all residents of the United States are green and have tentacles.

   d. *If* . . . all dogs are four-legged animals, and
     if . . . this animal is four-legged,
     then . . . this animal is a dog.

   e. *If* . . . all Martians are residents of the United States, and
     if . . . all humans are residents of the United States,
     then . . . all Martians are humans.

2.6. Think of something in your personal life that might represent a paradigm shift, for example, switching from a typewriter to a computer, learning a new language, or recognizing that a personal relationship is not going to work out. Describe some of the changing feelings, realizations, and behaviors you encounter during this shift.

2.7. P.F. and M.S. Klopfer of Duke University have studied maternal behavior in goats. The following facts have been established: A mother goat (doe) will reject her young (kid) if deprived of it immediately after birth, even if it is given back an hour later. If allowed contact with her own kid for 5 minutes after birth and then separated from it, the doe immediately accepts the kid and its littermates, if any, when returned 1 hour later. If allowed contact with her kid immediately after birth and then deprived of it, the doe shows obvious signs of distress. However, if this early contact is denied, the doe acts as if she had never mated or given birth to young.

    a. Propose a hypothesis to account for these observations, along with an experiment to test your hypothesis. Do not read the remainder of the exercise until you have finished this part.

Now consider the following additional facts about maternal behavior in goats established by the Klopfers: Under the conditions described above, the doe will not accept a kid of the correct age that is not her own (an alien kid) if it, instead of her own kid, is returned to her. Denied her own kid immediately after birth, but allowed 5 minutes with an alien kid, the doe not only will accept the alien kid but also her own when returned. However, only the alien kid with which she is allowed contact will be accepted; all other alien kids are rejected.

    b. How do these facts affect your hypothesis? (Do not change your hypothesis as given in the previous question, even if it did not fare too well— only consistency with the facts given and experimental design are of primary importance.)

    c. If necessary, propose a new hypothesis to account for all the Klopfers' data as given above and suggest an experiment to test this hypothesis. If your original hypothesis stands, fine. If it needs modifying, do so.

2.8. Much of science involves the use of intuition. Yet intuition is often deceiving. Hence scientists recognize that intuitive ideas must always be checked against empirical evidence. Consider the following problem. Imagine that you have ten marbles, identical in size and weight, of which eight are white and two are red. Now suppose you put these marbles into a brown paper bag and shake them up to randomize the system. Without peeking, you then reach into the bag and draw out a marble. If the marble is white, you put it aside and draw another. If that one, too, is white you put it with the other white marble you obtained on the first draw. You keep doing this until you draw a red marble. When that happens, you record the number of the draw on which you obtained the red marble—for example, the sixth draw. You then put all of the marbles you have drawn, both red and white, back into the bag with those remaining there and repeat the process 100 times or, better, 1,000 times.

The question is: On which draw are you most likely to get a red marble?

Further Reading

Farely, John, and Gerald L. Geison. Science, politics and spontaneous generation in nineteenth-century France: the Pasteur–Pouchet debate. *Bulletin of the History of Medicine* 1974;48(summer):161–198.
A thorough discussion of the political and religious context of the debate on spontaneous generation.

Grinnell, Frederick. *The Scientific Attitude*. Boulder, CO: Westview Press; 1987.
Written by a practicing scientist for undergraduates and graduate students in science, this book is a simple, straightforward introduction to many aspects of science as a process. Topics include problems of observation, experimental design and interpretation, science as a collective activity and "thought style" or how scientific ideas are perpetuated and become entrenched.

Longino, Helen E. *Science as Social Knowledge: Values and Objectivity in Scientific Inquiry*. Princeton, NJ: Princeton University Press; 1990.
In a clear introduction to problems of science as a social process, the author steers a sound course between the stereotype of science as objective truth and the view that it is nothing but subjective social construction. Deals with such issues as sex bias in research, the nature of evidence, values in science, and science as social knowledge.

Mayr, Ernst. Cause and effect in biology. *Science* 1961;134:1502–1506.
An elaboration of the example presented here on the causes of bird migration.

Temin, HM. The participation of DNA in Rous sarcoma virus production. *Virology*. 1964;23:486–494.
The original scientific presentation on Temin's provirus hypothesis.

Varmus, Harold. Reverse transcription. *Scientific American* 1987:257 (Sept): 56–64.
A detailed account of the discovery and process of reverse transcription and the action of the enzyme reverse transcriptase, as originally postulated by Howard Temin.

Suggested Websites

1. What is Science?
http://www.wsu.edu:8080/~dee/SCIENCE/SCIENCE.HTM
Part of the World Civilizations Distance-Learning Program at Washington State University, this site provides historical context for the development of science, from the Pre-Socratics through the European Enlightenment.

2. 2Think

http://www.2think.org/
> *Essentially an annotated list of book reviews concerning the scientific way of know-ing, this freethinking site argues consistently that we are responsible for learning about the world as it is, not as other would have us believe.*

3. Access Excellence Classic Collection

http://www.accessexcellence.com/AE/AEC/CC/
> *Each item in this collection is a case study of a single scientific event (including the discovery of penicillin and the development of polio vaccines). Included in each unit is a set of suggested references, activities, and graphics. Originally developed by Genetech, this site is now maintained by The National Health Museum.*

4. On Spontaneous Generation

http://guava.phil.lehigh.edu/spon.htm
> *The complete text of an address by Louis Pasteur to the "Sorbing Scientific Soiree" on April 7, 1864.*

5. Paul Kammerer and the Suspect Siphons

http://www.mbl.edu/publications/Ciona/Kammerer/
> *Noted embryologist J.R.Whitaker weighs in on another of Paul Kammerer's suspi-cious reports purporting to provide scientific evidence for the inheritance of acquired characteristics.*

6. The Historical Roots of Developmental Biology

http://www.ucalgary.ca/UofC/eduweb/virtualembryo/roots.html
> *Part of a larger site called* Developmental Dynamics, *this page discusses the Roux Driesch debate over vitalism and mechanism as competing explanations for embry-onic development. The value of this site is that sections are referenced to several col-lege textbooks in developmental biology.*

7. Thomas Kuhn

http://www.emory.edu/EDUCATION/mfp/Kuhnsnap.html
> *The most valuable aspect of site dedicated to the memory of Thomas Kuhn is the extensive list of links to essays about Kuhn's work.*

The Nature and Logic of Science: Testing Hypotheses

No matter how brilliant, a hypothesis that cannot be tested is of little use in science. Scientists test hypotheses in two different ways: by making further observations of the phenomenon under consideration and by carrying out experiments. We will examine both of these methods in this chapter.

Testing Hypotheses: Some General Principles

Testing Hypotheses by Observation

In many scientific investigations, the only test for a given hypothesis is to make additional observations. A paleontologist cannot go back in time to recreate the sequence of stages in the evolution of present-day species. Instead, he or she must infer information from already-existing data or make additional observations of the fossil record.

One problem with collecting new observations as the major means of testing a hypothesis is that it is not always possible to make the desired observations. For example, if a paleontologist cannot find fossil predecessors of mammals, any hypotheses about their evolutionary origin will have to remain untested. Sometimes we must simply wait for the right circumstances to present themselves, such as chance erosion of strata that exposes the needed fossils. Whenever possible, scientists prefer to test hypotheses by conducting experiments.

Testing Hypotheses by Experiments

Experiments are direct interventions into nature that allow the investigator to force specific situations to occur that might occur only rarely, if at all, on their own. Experiments make it possible to control the conditions under which a given effect occurs and thus allow the investigator to determine cause-and-effect relationships more precisely.

Suppose, for example, we wish to see whether light affects the growth of plant seedlings. We could roam the woods looking for seedlings of the same species that are germinating in the dark, perhaps under logs or under rocky ledges, and compare them with seedlings growing in an open field. Yet, under such conditions, rate of growth might be affected by many additional factors or variables, for example, differences in water or light received by the two groups of seedlings or differences in temperature. A more reliable answer to our question can be obtained by conducting an experiment in the laboratory, similar to one you may have carried out in elementary school. We might start by refining our question according to the information about meaningful scientific questions offered in Chapter 2. The broad question "Does light affect the growth of plant seedlings?" might be posed more specifically as:

1. Do plant seedlings need light to grow at all? or;
2. Do plant seedlings need light for normal growth?

The first is a simpler question, whereas the latter is more complex. For example, if we asked the second question, we would have to be prepared to know what was meant by "normal growth." In fact, we can answer both questions with one experiment, but the important point here is that, in framing the question initially, we need to think through what sorts of information we want to get from an experiment to answer the question posed.

If we formulate our question more broadly as "Do plant seedlings need light for normal growth?" then we can formulate a hypothesis that can be tested:

Hypothesis: *If* . . . plant seedlings do not require light for normal growth,

Prediction: *then* . . . seedlings grown in the dark should show the same amount of growth as those grown in the light.

An appropriate experiment to test this hypothesis would be as follows: Expose one group of seedlings to light and another group of the same species to total darkness for a specified period of time. The experimental design and results of such an experiment are shown in Figure 3.1. Recall from Chapter 2 that, in setting up experiments, the group in which the variable (in this case light) is altered is called the experimental group and the group in which the variable is not altered (because seedlings normally develop in light) is called the control group. It would also be important to

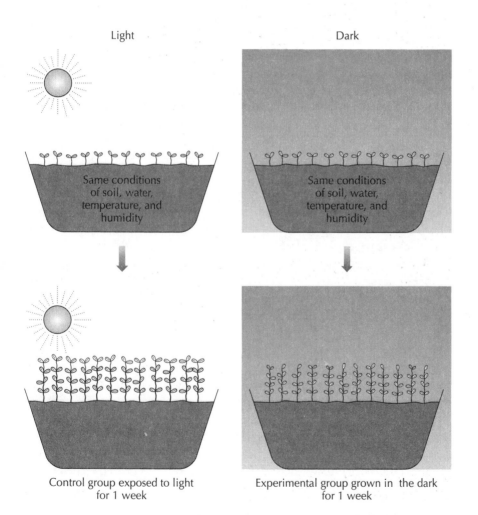

Figure 3.1 Controlled experiment. Plant seedlings grown in light and dark to test whether light is necessary for normal growth.

insure that all other known variables (temperature, availability of water, etc.) are kept the same for both groups. Observing the two groups after a week, we might notice that plants in the experimental group are lighter in color and thinner than the controls grown in the light, which, by comparison, are green and robust. In this experiment, the data contradict the hypothesis proposing that light is not required for normal growth. Thus, we can conclude that light *is* necessary for normal seedling development. Because experiments allow us to manipulate external variables, we can draw more reliable conclusions from them than from observations alone.

Note that we expressed the hypothesis as a negative: "If plant seedlings *do not* require light for normal growth...." We could just as easily have phrased the hypothesis as a positive: "If plant seedlings *do* require light for normal growth ...." Although both forms are acceptable, the negative form of stating a hypothesis has one logical advantage. Recall from Chapter 2 that to disprove a hypothesis is more certain than simply to support one. Researchers often phrase their hypotheses in the negative, or null form, stating that there will be no effect of the factor (in this case, light) being investigated. The null hypothesis is analogous to the legal principle of presumed innocence until proof of guilt. Because the results in our experiment above contradict the prediction based on the null hypothesis, we can reject it as disproved and conclude that the alternative, that light *is* necessary for normal growth, is at the very least highly probable. The null hypothesis is particularly useful in statistical tests applied to many experimental results. These tests determine whether the difference observed between experimental and control groups is statistically significant or merely to be expected by chance alone (see Appendix 1). Whenever possible, then, it is preferable to state hypotheses in the null form.

The Importance of Uniformity and Sample Size

It is not always possible to set up perfectly controlled experiments. Sometimes conditions cannot be controlled as rigorously as the investigator would like. As anyone who has tried to grow house plants knows, different species of plants may have very different requirements for water, humidity, light, or temperature. Indeed, considerable variations are often found among individuals *within* species. Thus, laboratory research is often carried out on organisms, such as mice or rats, that are inbred to be as biologically similar as possible. Another problem is that sometimes the organisms being studied, for example large mammals or members of an endangered species, cannot be obtained in sufficient quantities to make a large-scale study possible. In our seedling experiment, the availability of seedlings was not a problem, but, clearly, one or two seedlings would not have been enough from which to draw any conclusions. Only by having a large number of plants that are as biologically uniform as possible could we minimize the effects of individual differences. Even then, we obviously could not assume that this conclusion is valid for all plants but, instead, is valid only for the species with which we worked. As discussed in Chapter 2, Spallanzani recognized this issue in drawing his conclusions.

The problem of adequate controls is made more complex when the organisms under consideration are human beings. Ethical issues usually prevent investigators from performing fully controlled laboratory studies, although investigators do their best within certain limits. If we hypothesized, for example, that substance X causes cancer, we could not ethically or legally divide a group of humans into experimental and control groups, administering X to one group and not the other. This is not to say, however, that scientific procedure has no role to play in studying the causes of human diseases or other issues of human biology. As noted in Chapter 2, clinical trials are common in medical studies among designated populations, such as smokers or patients with acquired immunodeficiency syndrome (AIDS), and in which the individuals in the study are both (1) volunteers carefully selected for the study and (2) persons for whom the results of the study might provide a corrective or cure to their habit or condition. Here, the ethical problem of asking a person to volunteer for an experimental study must be weighed against the chance that participation might benefit them directly. The question always exists: What is a true volunteer in any given case? Is a patient suffering from a terminal illness truly volunteering or acting out of desperation? Can a member of a prison population, to whom certain privileges are granted for participation in a medical study, truly volunteers? Are elementary school children, whose parents are paid a small fee to test behavior-controlling drugs, truly volunteers?

Even if the ethical questions are resolved, clinical trials involving humans face other methodological problems, among which finding adequate controls is probably the most important. Suppose we want to investigate whether drug Y is effective in alleviating the effects of allergies. First, we must match an experimental and a control group for all sorts of factors that might influence interpretation of the results. Allergies are influenced by many environmental factors, including geographic locality, climate, drinking water supply, diet, lifestyle, as well as age, state of health, and socioeconomic status. It is obvious that the effectiveness of a drug cannot be judged if it is given to an experimental group with an average age of 35 and few smokers, and the control group is composed of people with an average age of 55 and many smokers. Finding comparable sample populations is one of the most difficult problems in constructing and carrying out clinical research.

However, even if all the above factors are matched between control and experimental groups, another problem remains in clinical studies that the laboratory experimenter does not face: Humans cannot be kept in a controlled environment for long periods of time. In most cases, the individuals involved in such studies carry on routine daily activities that may differ considerably from one person to the next. We might imagine the comparable situation for a laboratory investigator who opens the cages every night and lets the experimental and control mice run wild outside and then re-collects them in the morning. The investigator working with human populations can only hope that the variation in activity and lifestyles among subjects is not so great as to undermine the validity of any results obtained.

A Case Study in Hypothesis Formulation and Testing: How is Cholera Transmitted?

Cholera is a highly contagious and often fatal disease. Its initial symptoms are severe diarrhea, nausea, and vomiting, followed by painful muscle cramps, eventually leading to extreme dehydration. In the past several centuries, cholera epidemics have swept Asia, western Europe, Africa, and North America. Before the development of the germ theory of disease by Louis Pasteur and Robert Koch (1843–1910) in the late nineteenth century, neither the cause of cholera nor its means of transmission were understood.

Disease: The Early Views

The Greek physician Hippocrates (460–370 BCE) argued that disease was the result of environmental factors, such as polluted soil, water, or air. In the intervening centuries, many cultures turned to supernatural rather than natural explanations for disease; for example, that an epidemic or an individual's illness was the result of God's anger. Even today, this view occasionally surfaces—for example some television evangelists have attributed AIDS (see Chapter 4) to God's punishment for the "sin" of homosexuality.

It was within a similar atmosphere that British physician John Snow (1813–1858) practiced medicine in the mid-nineteenth century. Although Snow is probably best known for being among the first to administer chloroform as an anesthetic during childbirth (twice to Queen Victoria), his greatest contribution to medicine lay in establishing the means of cholera transmission. Snow's investigations stemmed from two cholera outbreaks in London, the first in 1848 and 1849 and the second in 1853 and 1854. In 1854, Snow published his findings in a monograph entitled *On the Mode of Communication of Cholera*. Snow's work, which led ultimately to a method for controlling and preventing cholera epidemics (effective most recently in the February 2000 floods in Mozambique), is a simple yet elegant example of scientific investigation.

The main problem facing Snow was: How is cholera transmitted? Although microscopic organisms (or "animacules," as they were then called) had been known to exist since the seventeenth century, solid clinical and experimental support for the germ theory of disease would not begin to be accepted until the 1870s. Thus, the role of bacteria in the transmission of communicable disease was unknown during Snow's investigations of cholera. The most prevalent view at the time was the "effluvia" hypothesis, which held that cholera was transmitted by poisoned air exhaled by cholera patients or emanating from the corpses of those who had died of the disease.

Snow's Observations

Snow had made a number of observations about cholera that led him to doubt the effluvia hypothesis. He first recorded "certain circumstances" associated with the spread of cholera:

> It travels along the great tracks of human intercourse [interaction] never going faster than people travel, and generally much more slowly. In extending to a fresh island or continent, it always appears first at a seaport. It never attacks the crews of ships going from a country free from cholera, to one where the disease is prevailing, till they have entered a port, or have [interacted] with [people on] the shore. [This and other 1854 quotations from Snow are taken from *Snow on Cholera*. New York, NY: Commonwealth Fund; 1936. Bracketed inserts in the quotations have been added for clarification.]

Snow next described several cases of the disease with which he had personal experience and that, to him, cast some doubts on the effluvia hypothesis:

> I called lately to inquire respecting the death of Mrs. Gore, the wife of a labourer, from cholera....I found that a son of the deceased had been living and working at Chelsea. He came home ill with a bowel complaint, of which he died in a day or two. His death took place on August 18th. His mother, who attended on him, was taken ill on the next day, and died the day following (August 20th). There were no other deaths from cholera registered in any of the metropolitan districts, down to the 26th August, within two or three miles of the above place....

John Barnes, aged 39, an agricultural labourer, became severely indisposed on the 28th of December 1832; he had been suffering from diarrhea and cramps for two days previously. He was visited by Mr. George Hopps, a respectable surgeon at Redhousek, who...at once recognized the case as one of Asiatic cholera....

While the surgeons were vainly endeavouring to discover whence the disease could possibly have arisen, the mystery was all at once, and most unexpectedly, unraveled by the arrival in the village of the son of the deceased John Barnes. This young man was apprentice to his uncle, a shoemaker, living at Leeds. He informed the surgeons that his uncle's wife (his father's sister) had died of cholera a fortnight before that time, and that as she had no children, her wearing apparel had been sent to [John Barnes] by a common carrier. The clothes had not been washed; Barnes had opened the box in the evening; the next day he had fallen sick of the disease.

For Snow, the implication of these observations was that cholera seemed to be passed from a sick to a healthy person by some sort of direct contact between individuals, that is, by some material agent that could be transmitted even by way of a patient's clothes. It seemed clear that John Barnes had not been in his sister's presence during her illness and, hence, could not have come in contact with any sort of poisoned effluvia. However, he had handled her unwashed clothing. After pointing out that the medical literature was full of additional cases similar to the ones he cited, Snow concluded that:

...the above instances are quite sufficient to show that cholera can be communicated from the sick to the healthy; for it is quite impossible that even a tenth part of these cases of consecutive illness could have followed each other by mere coincidence, without being connected as cause and effect....

Snow made yet another observation. Many people, such as doctors and nurses who dealt regularly with cholera patients and thus should have been constantly exposed to effluvia, usually did not become afflicted with the disease. He also noted that doctors usually washed their hands routinely after handling patients or their clothes. We might formulate Snow's doubts about the effluvia hypothesis deductively:

Hypothesis: *If* . . . cholera is spread by effluvia emanating from the body of an afflicted individual, and healthy individuals are exposed to these effluvia,

Prediction: *then* . . . all persons in close contact with effluvia emanating from afflicted patients or their bodies should get cholera.

An Alternative Hypothesis

Snow further observed that cholera begins with intestinal symptoms rather than the skin lesions, coughing, or respiratory ailments that the effluvia hypothesis would have predicted. To Snow, this observation implied that the infectious agent must be ingested through the mouth, from which it then entered the digestive system directly. He therefore proposed an alternative to the effluvia hypothesis. Noting that the clothes and bed linens of cholera patients

> ...nearly always become wetted by the cholera evacuations, and as these are devoid of the usual colour and odour, the hands of persons waiting on the patient become soiled without their knowing it; and unless these persons are scrupulously clean in their habits, and wash their hands before taking food, they must accidentally swallow some of the excretion, and leave some on the food they handle or prepare...

Snow was clearly moving toward a hypothesis proposing that the agent causing cholera was transmitted primarily, if not exclusively, from infected to healthy persons by ingestion of contaminated food or water. A further set of observations supported this view. Noting that cholera was particularly prevalent among the "working classes," a term often used during the Industrial Revolution for urban laborers who lived crowded together in vast slums, Snow made the following observations:

> Mr. Baker, of Staines, who attended 260 cases of cholera and diarrhea in 1849, chiefly among the poor informed me...that where a whole family live, sleep, cook, eat and wash in a single room, that cholera has been found to spread when once introduced, and still more in those places termed common lodging-houses, in which several families were crowded into a single room.

By contrast, Snow noted that:

> When, on the other hand, cholera is introduced into the better kind of houses...it hardly spreads from one member of the family to another. The constant use of the hand-basin and towel, and the fact of the apartments for cooking and eating being distinct from the sick room, are the cause of this....

Because the poor, especially those with cholera, had little direct contact with the rich in Victorian England, how would the wealthy ever contract the disease, as some of them certainly did? Snow offered an explanation:

> ...there is often a way open for it [cholera] to extend itself more widely, and to reach the well-to-do classes of the community; I allude to the mix-

ture of the cholera evacuations with the water used for drinking and culinary purposes, either by permeating the ground, and getting into wells, or by running along channels and sewers into the rivers from which entire towns are sometimes supplied with water....

Snow noted that, especially in areas of poor sanitation like those in the London slums, the "cholera evacuations" might also get into the groundwater. He was now ready to advance a more general form of his hypothesis: Cholera was spread within a community not only by contact between individuals but through the local water supply. The sewers of London emptied into the Thames, which also supplied the city's drinking water (Fig. 3.2). Although there was no direct evidence that cholera was carried by water, additional observations provided indirect support for Snow's alternative hypothesis. Among his observations were:

1. In Manchester, England, Hope Street residents using water from one particular well were subjected to a severe cholera epidemic; 25 out of 26 perished from the disease. Residents in the same neighborhood who used water from another well showed no cholera.

Figure 3.2 "Monster Soup, commonly called Thames water," 1828 etching in Philadelphia Museum of Art. London residents were aware of the pollution of the Thames, although they they did not connect it with the spread of diseases like cholera.

Examination of the first well showed that a sewer which passed only 9 inches from it had leaked into the well.

2. In Essex, in 1849, cholera attacked every house in a row except one. This house turned out to be occupied by a washerwoman. She noted that the water (from the local well) gave her laundry a bad smell, and so used water from another part of town for washing, cooking, and drinking.

3. In Locksbrook, cholera appeared in several tenement houses. The tenants had complained about the water to the landlord, who lived in the nearby town of Bath. He sent a surveyor, who could find nothing wrong. The tenants continued complaining. This time the landlord came. He smelled the water and pronounced it fine. Challenged to try it, he drank a glassful. This was Wednesday; by Saturday he was dead. His was the only case of cholera in Bath.

The Case of the Broad Street Pump

In 1849, a severe outbreak of cholera occurred in the vicinity of Broad and Cambridge Streets in London. At that intersection, a pump supplied water to the neighborhood. Snow quickly noted that the epidemic appeared to be centered at the pump. To verify that impression, he mapped the occurrence of the disease house by house (Fig. 3.3). The results were startling. Within 250 yards of the pump, almost 300 fatalities occurred in 10 days, a result quite consistent with his hypothesis. However, Snow had to deal with what appeared to be a contradictory observation. Seventy workers in a Broad Street brewery, located very close to the pump, remained free of cholera. This contradiction was resolved, however, when Snow learned that, when the workers were thirsty, they drank water from the brewery's own deep well rather than from the Broad Street pump. On September 7, Snow convinced a citizens' group of the danger of the Broad Street pump. On September 8, its handle was removed. However, the epidemic already had reached its peak, the number of deaths had begun to decline, and most of the inhabitants had fled the area. Thus, removal of the pump handle came too late to be viewed as a means of testing Snow's water-borne hypothesis.

Objections to Snow's Water-Borne Hypothesis

Several objections were raised to Snow's water-borne hypothesis. Some people argued that not everyone who drank the polluted water became ill, precisely the same objection Snow himself had used against the effluvia hypothesis. Snow's reply made two important distinctions that showed considerable sophistication on his part.

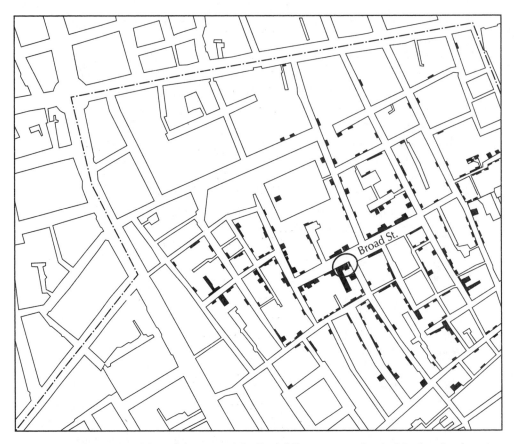

Figure 3.3 Map of the area of the Broad Street pump (in circle), showing location of cholera cases, 1849. The clustering of cases near the pump first alerted Snow to the possible contamination of water from this well by the cholera agent.

The first distinction was that between a simple and complex cause-and-effect relationship. For his water-borne hypothesis to be supported, it was not necessary that every person who drank from the same contaminated well come down with the disease. For various reasons, including the state of health of the individual or the amount of water consumed, not all individuals would be expected to respond identically.

The second distinction was between what Snow called a "chemical" and a "biological" cause. Here, his reasoning is particularly insightful. *If* the infective agent of cholera were simply a chemical, Snow noted, *then* it would tend to be diluted in drinking water after a time and thus would not persist with continued virulence for weeks at a time. On the other hand, *if* the infecting agent were a biological one capable of reproducing itself, *then* it would not be diluted over time but spread over wider and wider areas. Clearly, he remarked, the pattern of infection showed that as long as people were in contact with one another, the cholera spread rapidly and showed no signs of becoming diluted. Therefore, Snow reasoned, it must be a biological rather than a chemical agent.

The Critical Test

How could Snow test his hypothesis in a way that would distinguish clearly between the two alternatives? As we noted earlier, it is difficult and often unethical to carry out controlled experiments with human beings, especially if the point is to determine susceptibility to a fatal disease. However, in 1853, another cholera epidemic in London provided the experiment that allowed Snow to test his water-borne hypothesis directly.

Two water companies, the Lambeth Company and the Southwark and Vauxhall Company, both delivered water to a single district in London. As a consequence of the 1849 cholera epidemic, however, the Lambeth Company moved its waterworks upstream from the city, where the Thames was free of city sewage. The Southwark and Vauxhall Company continued to take water downstream. This situation provided the opportunity for a critical test of Snow's hypothesis. As Snow reported it:

> Each Company supplies both rich and poor, both large houses and small; there is no difference either in the condition or occupation of the persons receiving the water of the different Companies....it is obvious that no experiment could have been devised which would more thoroughly test the effect of water supply on the progress of cholera than this....The experiment, too, was on the grandest scale. No fewer than 300,000 people of both sexes, of every age and occupation, and of every rank and station, from gentlefolks down to the very poor, were divided into two groups without their choice, and, in most cases, without their knowledge; one group being supplied with water containing the sewage of London....the other group having water quite free from such impurity.

Stated deductively in the null form, Snow's hypothesis predicted that *if* cholera were not borne by water, *then* there should be no difference in cases of cholera between those houses receiving Southwark and Vauxhall water and those receiving water from the Lambeth Company. His preliminary data determined that, of 44 deaths from cholera in the district, 38 occurred in houses supplied by Southwark and Vauxhall. The null hypothesis was contradicted.

What about the other six cases? Although most of the data supported Snow's water-borne hypothesis, he recognized that the six cases represented a false prediction and thus a possible negation of his hypothesis. Rather than abandon his hypothesis, however, Snow did what most scientists do. He tried to find variables that might explain the six exceptions. He found that, in many cases, the city residents (especially tenants whose landlords paid the water bill) did not know which water company supplied their house. How could he correct for this possible source of error? Snow found that he could distinguish the source of the water by a simple chemical test based on the fact that the two water supplies possessed different levels of sodium chloride. The difference in the amount of sodium chloride in the two water supplies was great enough to provide a reliable and objective way for Snow to determine which company supplied water to each house. His results showed that all six exceptions actually received their water from Southwark and Vauxhall, precisely as would be predicted by the water-borne hypothesis.

Note that Snow's use of data from the two water companies allowed him to control for other variables that might have affected spread of the disease. As Snow noted, "Each company supplies both rich and poor, both large houses and small; there is no difference either in the condition or occupation of the persons receiving the water of the different companies...." Thus all the various social and economic factors that might have influenced the spread of the disease were eliminated as causal factors as thoroughly as they could be in such a large-scale and naturally occurring experiment. This left the one major factor—polluted or clean water—as the single difference between two large groups of subjects. By chance, Snow had been able to devise a controlled experiment without actually having to intervene directly and divide his subjects into experimental and control groups.

A Wider Applicability and the Role of Chance

Snow's work on cholera demonstrates two other factors often characteristic of scientific investigations. One is that hypotheses often apply to more than the phenomenon they were originally designed to explain. For example, Snow noted that typhoid fever killed even more people than did cholera, suggesting that typhoid fever might also be corrected by improved sanitation. This turned out to be the case, and typhoid fever, too, eventually was brought under control. Another factor is that new ideas often emerge in scientific investigations but are often overlooked. In his strong focus on the question of how cholera was transmitted, Snow missed an obvious therapeutic treatment coming directly from his own work. In making his case for the oral transmission of cholera, Snow wrote:

If any further proof were wanting [that] all the symptoms attending cholera, except those connected with the alimentary canal, depend simply on the physical alteration of the blood and not on any cholera poison circulating in the system, it would only be necessary to allude to the effects of a weak saline solution injected into the veins [of a patient] in the stage of collapse. The shrunken skin becomes filled out, and loses its coldness and lividity; the countenance assumes a natural aspect; the patient is able to sit up, and for a time seems well. If the symptoms were caused by a poison circulating in the blood depressing the action of the heart, it is impossible that they should thus be suspended by an injection of warm water, holding a little carbonate of soda in solution.

Note the deductive reasoning here: *If* the symptoms were caused by a poison circulating in the blood..., *then* they should not be alleviated by a simple injection of warm water. Snow overlooked here a potentially powerful treatment for cholera. We know today that the disease kills not by the effect of bacterial toxins directly but by dehydration. If a cholera patient drinks enough fluids or if fluids are given intravenously, as Snow described, the disease is rarely fatal. It is ironic that Snow was so focused on finding the cause of cholera that he overlooked a possible treatment, even when it was directly in front of him.

The Mechanism of Cholera Action: A Modern Perspective

Modern-day studies on the nature of cholera infection have revealed how the bacterium *Vibrio cholerae* works its lethal effects on the human body. During the course of its metabolism, the bacterium produces a toxin that affects the ability of epithelial (lining) cells, such as those lining the inside of the intestines, to control the flow of water out of the cell. As a result of cholera infection, the individual loses so much water (thus the symptom of extreme diarrhea) that he or she becomes dehydrated and, if untreated, eventually dies. This modern understanding of the mechanism by which cholera operates now makes perfectly clear the reason behind Snow's observation that when cholera patients were injected with saline solution their condition improved: The injection was re-hydrating them. One therapy routinely applied today for cholera patients is to give them massive injections of water, thus keeping them alive so that the body's immune system has time to fight off the infection.

A Modern Epidemic: The AIDS Crisis, 1981–?

Background: The Origin of the AIDS Crisis

What cholera was to the nineteenth century, AIDS has become for the twentieth. AIDS is a particularly deadly disease. It involves a breakdown of the body's immune system, which has the job of warding off infection and constantly surveying the body's tissues for cancerous cells. People with AIDS do not die directly from the disease but from one or more of any number of infectious diseases or cancers that healthy people easily ward off. Epidemiological data from the Centers for Disease Control and Prevention in Atlanta show that, in the United States, the number of reported cases of AIDS went from less than 100 before 1981 to more than 79,000 per year in the peak year spanning 1992 and 1993 (Fig. 3.4). Although the rate of increase began to fall off by the mid-1990s in the United States and other industrialized nations, it was simultaneously escalating in the rest of the world, especially in Africa and Asia. In 1997, nearly 6 million people came down with AIDS worldwide, and more than 2 million died of the disease, at least 460,000 of them children. AIDS has moved from epidemic (localized outbreaks) to pandemic (affecting whole continents) proportions. Today, when humans have such a high degree of mobility, the spread of disease can occur much more rapidly and extensively than at any other time in history. The causes of AIDS and methods of treating it have challenged medical science and public health policies in unprecedented ways.

What causes AIDS, and how can scientists help fight back against this deadly disease? Since it made its first official appearance as a recognizable

Number of new AIDS cases reported

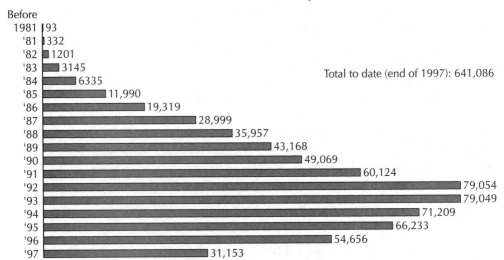

Figure 3.4 Graph showing increasing incidence of AIDS in the United States, 1981–1997. The noticeable dropoff after 1995 reflects the use of new drug combinations. Single drugs had been common as treatment until the mid-1990s. Statistics from the Centers for Disease Control and Prevention, Atlanta, GA

disease in 1981, AIDS has been the source of controversy regarding both origin and treatment. Some of the basic observational data regarding AIDS are:

1. No one comes down with AIDS without being in contact with someone who already has the virus believed to cause the disease. Like cholera, AIDS meets all the criteria of a transmissible disease.

2. Although AIDS is transmissible, it not *easily* transmissible. It cannot be caught by simple physical contact, such as touching or kissing. Its only major portal into the human body is by way of the blood stream. AIDS is most often thought of as a sexually transmitted disease because, during sexual intercourse, capillaries in epithelial tissue lining the reproductive tract or other body parts may be torn, providing a direct passageway into the blood stream. However, the disease may also be transmitted among intravenous drug users who use nonsterile needles and, very rarely, by the medical transfusion of blood and blood-derived products or by accidental needle sticks in the hospital setting.

3. What came to be called AIDS was identified first in western industrialized countries within the homosexual community, where it became highly visible (for a short while it was called the "gay disease"). AIDS later showed up in the heterosexual community and among drug users and prostitutes in both industrialized and nonindustrialized countries.

4. People who have sexual or intravenous contact with infected individuals and who acquire the virus that causes AIDS from this contact often experience a long latent period of anywhere from a few months to 10 years or more, during which time they show no signs of the disease.

5. People who have contracted AIDS show a reduction in the number of CD4+ T cells (also called helper T cells), components of the human immune system that recognize foreign invading elements and mobilize other parts of the immune system to eliminate the invaders before they do significant harm. Individuals with AIDS therefore lack this first line of defense against a whole host of infective agents.

6. People who show clinical symptoms of AIDS also have elevated levels of a virus know as human immunodeficiency virus (HIV) in their blood streams.

7. Until 1995, anyone who began to experience the overt symptoms of AIDS was likely to be dead within 2 to 3 years. Since 1995, however, various combinations of drugs have delayed this previously inevitable fate. There is some hope now that, even if AIDS cannot be cured, it can at least be managed so that people with the disease can live a relatively normal and extended life.

To cure or control AIDS effectively, medical and public health officials sought very hard in the period after 1981 to determine the cause of the disease. Three general explanatory hypotheses have been advanced since that time to account for the AIDS pandemic:

1. AIDS is God's punishment for homosexuality or intravenous drug use. Representatives of the religious right, including various radio talk show hosts and some television evangelists, have promoted this hypothesis.
2. AIDS is a result of a degenerate lifestyle that involves compromising the immune system through excessive sex, drug (especially amyl nitrate, inhaled as "poppers") and alcohol use, lack of proper diet, etc. Infective agents associated with AIDS are thus "fellow travelers" of an unhealthy lifestyle and not the direct cause of the disease. Dr. Peter Duesberg, a biochemist at the University of California, Berkeley, is one of the major proponents of this view.
3. AIDS is caused by infection from the HIV virus, which specifically attacks (enters) CD4+ T cells. CD4+ cells are killed as the virus replicates itself inside them and, as the virus spreads throughout the entire population of CD4+ T cells, the body's whole immune system is slowly rendered ineffective

It is possible to distinguish between these alternative explanations. Hypothesis 1 can be eliminated immediately, because it postulates a mystical, supernatural cause and therefore cannot be tested. Hypotheses 2 and 3, on the other hand, may be distinguished from one another by experimental and observational means. For example:

If . . .   AIDS is the result of a harmful lifestyle (rather than a direct effect of HIV infection),

then . . .   people who live a healthy lifestyle should not come down with AIDS less frequently than those who lead an unhealthy lifestyle.

Medical records show, however, that health care workers and others who have led perfectly healthy lifestyles can contract the disease if they experienced needle punctures while caring for AIDS patients. A healthy lifestyle, therefore, does not appear to be a guarantee against contracting AIDS. The prediction was not borne out, thus weakening hypothesis 2. A more direct prediction can be made from hypothesis 3:

If . . .   AIDS is the direct result of infection by the HIV virus,

then . . .   infecting cells in laboratory cultures or whole organisms with HIV should produce AIDS-related effects.

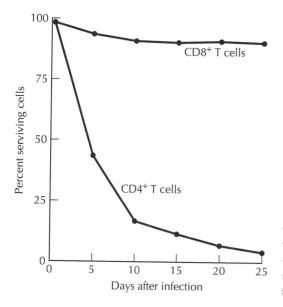

Figure 3.5 Graph of HIV infecting and destroying CD4+ T cells but not CD8+ T cells. These data show the specificity of the virus for certain cells and not others in the immune system.

Laboratory experiments show that the HIV virus readily infects CD4+ T cells in vitro (that is, in a culture dish) and destroys them, whereas it does not affect other kinds of T cells (Fig. 3.5). Furthermore, an animal model of AIDS provides a test for hypothesis 3. Monkeys carry related viruses known as simian immunodeficiency viruses (SIV). When SIV is injected into healthy monkeys, they soon begin to experience an AIDS-like immune response. (Monkeys, however, do not usually progress to having the disease.) By all counts, then, hypothesis 3 seems to be most strongly supported: AIDS appears to be directly caused by infection from the HIV virus.

Where did AIDS come from, and how did it make its appearance in human populations? Again, several hypotheses explain the rather recent and sudden appearance of the disease:

1. HIV is a cousin of SIV and evolved from SIV that had been transmitted to human populations through contaminated polio vaccines produced from infected monkey kidney cells. A widespread polio vaccination campaign was waged in central Africa in the 1950s, and it is postulated that an early form of HIV began to enter the human population at that time, becoming visible as an epidemic only a generation (30 years) later. This hypothesis has been championed most recently by Edward Hooper in a 1999 book, *The River: A Journey to the Source of HIV and AIDS.*

2. HIV was transmitted from African chimpanzees to humans, probably through bites or through hunting of chimps for food, with the virus adapting to exploit the human immune system.

Because both hypotheses rest on circumstantial evidence, it is not easy to distinguish between them. Hypothesis 1 may be true, but it is difficult to determine whether any of the polio vaccines produced in the 1950s involved monkey kidney cell cultures that were contaminated with HIV. Hypothesis 2 seems more likely to be correct at the present time, but we may have to be reconciled to the fact that it is difficult to determine the true origin of this disease. (Another hypothesis proposes that HIV originated in Caribbean pigs eaten by tourists in Haiti in the 1980s, and another suggests that HIV is the result of a military germ warfare program gone awry. Neither hypothesis has been supported by evidence, however.)

The Nature of HIV Infection

The front line of the body's defense system against invasion by foreign agents (principally bacteria, viruses, and fungi) are several types of cells of the immune system, including CD4+ helper T cells. Helper T cells respond to the presence of foreign bodies in the blood and body tissues by activating other types of cells of the immune system (B cells and cytotoxic T cells) that attack the invader. HIV viruses specifically infect the helper T cells. The HIV virus is able to target these specific cells because it is covered by a lipid envelope (lipid molecules are a major component of fats) in which are embedded glycoprotein markers. The glycoprotein markers recognize cell surface molecules on CD4+ T cells and bind to the cell surface. This specificity is indicated by the graph in Figure 3.5. In the course of 3 weeks after infection by HIV, virtually 100% of the CD4+ T cells in a laboratory culture have been eliminated from the body, whereas another kind of T cell, CD8+ T cells, remain virtually unaffected. Once the HIV virus particle attaches itself to a specific cell surface marker, the process of infection begins.

Like all viruses, HIV is very simple in structure. The virus consists of its lipid envelope (some viruses do not have a lipid envelope and have only a protein coat), inside of which is a capsid containing two long molecules of nucleic acid (the virus's genetic information system for making new viruses) and two molecules of the enzyme reverse transcriptase (Fig. 3.6). The envelope serves two purposes in the viral life cycle. It protects the nucleic acid from being destroyed by other molecules in the body or in the outside environment, and the glycoprotein fits specifically onto the surface molecules of T cells, initiating the process of infection.

The infection cycle of HIV begins with attachment of the virus to the T cell surface (Fig. 3.7A). Once attached, the virus injects its nucleic acid into the interior, and the work of infection begins (Fig. 3.7B). Infection involves the viral nucleic acid commandeering the machinery of the host

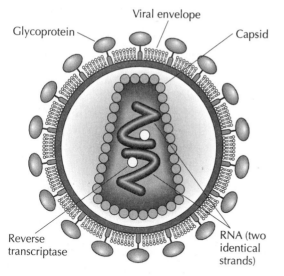

Viral envelope

Glycoprotein

Capsid

Reverse transcriptase

RNA (two identical strands)

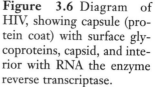

Figure 3.6 Diagram of HIV, showing capsule (protein coat) with surface glycoproteins, capsid, and interior with RNA the enzyme reverse transcriptase.

cell to produce not what the host cell needs but, instead, hundreds of new viral particles (Fig. 3.7C). The virus is like a saboteur who invades a factory and turns the production process toward the saboteur's own ends. Finally, the new virus particles are assembled and pass through the membrane of the destroyed helper T cell, each forming its new envelope from the lipid membrane of the cell (Fig. 3.7D). With each single virus being able to produce hundreds of progeny within one host cell, it is easy to see how the infection may spread rapidly throughout the body.

HIV is among a group of viruses known as retroviruses, with nucleic acid made of ribonucleic acid (RNA) rather than deoxyribonucleic acid (DNA) (see Chapter 2). Most viruses, like bacteria and all multicellular organisms, have DNA as their genetic material. Thus the host cell's metabolic machinery is set up to take instructions from DNA for whatever proteins and other molecules it synthesizes. Because they possess RNA as their genetic material, retroviruses must carry a specific enzyme, reverse transcriptase, that allows their RNA, once inside the cell, to first be transcribed into DNA. Once this transcription has taken place, the new DNA, carrying viral genetic information, can then provide instructions to the cell's machinery to produce viral RNA and proteins.

During the early period of infection, patients may be diagnosed as HIV-positive. Although the virus can be detected in their bodies, they show no other signs of illness. Medically speaking, they do not have AIDS. AIDS itself is said to occur when visible signs of the breakdown of the immune system are apparent. These signs can include repeated opportunistic infec-

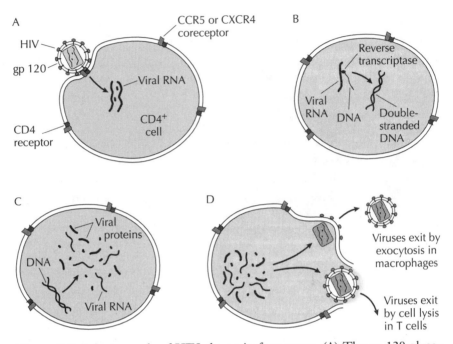

Figure 3.7 Infective cycle of HIV, shown in four stages. (A) The gp 120 glyco-protein on the surface of HIV attaches to CD4 and one of two co-receptors on the surface of a CD4+ cell. The viral contents enter the cell by endocytosis; (B) Reverse transcriptase makes a DNA copy of the viral RNA. The host cell then synthesizes a complementary strand of DNA; (C) The double-stranded DNA directs the synthesis of both HIV RNA and HIV proteins; and (D) Complete HIV particles are assembled in macrophages, HIV buds out of the cell by exocytosis. In T cells, however, HIV ruptures the cell, releasing free HIV. (Adapted from Peter Raven and George Johnson: *Biology, 5th ed.* St. Louis: MO: William C. Brown: 1999.)

tions by yeasts, the appearance of certain rare cancers associated with immune system deterioration, etc. Some individuals have remained HIV-positive for years without showing any major symptoms of AIDS.

For reasons that biologists still do not understand, HIV sometimes adopts a second strategy in its infectious process. Once it has transcribed DNA from its RNA inside the cell, the viral DNA becomes integrated into the chromosomes of the host, replicating as that cell replicates for many generations and producing a so-called latent period. This latent phase of the cycle (called the lysogenic phase, which means the virus has the potential to lyse [destroy] the cell but does not do so immediately) may account for the fact that people can be HIV-positive for many years before experiencing the symptoms of AIDS itself. Suddenly, however, for reasons equally mysterious, the viral DNA may become active and start the process of replicating HIV proteins and RNA.

The HIV virus is quite fragile. It cannot exist outside of a host cell or the blood stream for any length of time. In this sense, therefore, it should be a difficult disease to contract. Why, then, has AIDS become the major epidemic disease of the last two decades? The answer lies in several characteristics of this virus compared with most other viruses or bacteria with which medical practitioners have had to contend. The first is that, because HIV attacks the immune system, which is supposed to ward off such invasions, it paves the way for complete takeover of the body. The infected person becomes helpless, as his or her immune system breaks down. Second, HIV has the ability to mutate more rapidly than almost any other biological entity known. Mutations are particularly prominent in genes coding for the structure of the glycoprotein markers in the envelope. The immune system depends on being able to recognize invading microorganisms (including viruses) as foreign elements by detecting specific markers on their surfaces. If HIV strain 1 enters the body and the immune system learns to recognize it and eliminate it before it can reproduce, the strain will be kept at a low level or eliminated altogether. However, in the course of its replication within the body of a single host, HIV may undergo numerous mutations of its glycoproteins. As a result, after several generations of replication, it is no longer recognizable to the immune system as strain 1 and is therefore seen as a brand new invader. Antibodies now are made to this new HIV-strain 2, but, in the meantime, a number of T cells have been destroyed and the immune system weakened. This downward spiral is repeated until the entire immune system has collapsed.

AIDS has spread rapidly and on a global basis for social as well as biological reasons. Widespread increases in drug traffic (intravenous drug use is particularly culpable in spreading HIV), unregulated prostitution, and unsafe sex practices in general have contributed to the spread of this disease. Although the rate of increase in spread of the disease has diminished in industrialized countries, the situation is particularly tragic in impoverished areas of the Third World. In areas where drug trafficking and sex for money often appear as the only roads out of poverty or means of survival, the future is grim no matter what. As a prostitute in Harare, Zimbabwe, stated, "I can choose to die of starvation now or of AIDS later." Reports in October 2000 from the Thirteenth International AIDS Conference in South Africa produced some startling and disturbing facts: 15 million people have died of the disease worldwide, and another 34 million are HIV-positive, including 25 million in sub-Saharan Africa alone. It is projected that in some African countries 67% of today's teenagers will die of AIDS in the next decade. The available drug mixtures necessary to counter retroviral activity cost an average of $10,000 a year per person and can buy survival. In most Third World countries, however, the average annual expense on total health care is $80 per person. The gap between the rich and poor countries of the world makes the difference between life and death.

Treatment and/or Cure for AIDS

Until the mid-1990s, anyone diagnosed as HIV-positive was almost certain to die within a few years. The progress of the disease is horribly debilitating, with one infection following another, along with the appearance of rare cancers, loss of weight, and the breakdown of organs such as the kidney and liver. Various drugs have been tried, most notably zidovudine (usually called AZT), which inhibits the action of the enzyme reverse transcriptase. If this enzyme could be inhibited, it was hoped, viral replication would be halted at the outset. However, AZT alone produced some serious side effects, and it soon became clear that the virus mutates so quickly that its reverse transcriptase enzymes (coded by its own RNA) can evolve to circumvent the drug's effects. Thus AZT could retard the progress of the disease temporarily, but nothing seemed to work permanently. Hopes for a vaccine seemed even more remote. It appeared this was a virus that had a perfect strategy.

However, starting around 1995, clinicians began a new strategy known as "combination drug therapy." The idea here was to hit the virus hard at two different points in its replication cycle: once at the reverse transcriptase phase by using analogs of AZT, and again at the end of the replication cycle, with drugs known as protease inhibitors. A drug analog has a molecular structure that resembles that of the original drug but with sufficient differences so that side effects are reduced. AZT analogs proved to have fewer side effects than AZT itself. The protease inhibitors slow or stop the action of viral enzymes that cut into small pieces the viral protein produced during infection (these proteins serve as part of the protein coat and as enzymes such as reverse transcriptase). Together, these combination drugs have produced some remarkable effects. A combination of two AZT analogs and a protease inhibitor, for example, have lowered the level of HIV in many patients' blood below the level of detection. Some side effects, such as diarrhea, persist, and it is not clear whether the virus, with its high mutation rate, will evolve to circumvent these drugs. Yet the combination clearly has made it more difficult for the virus to mutate to resistant forms, and so, for the moment, it offers a way of staving off the disease while other strategies are explored.

One recent and exciting finding is that a few patients from Australia have been HIV-positive for 14 years and have never shown any signs of the disease. Isolation and study of the virus from these individuals has revealed that they are infected by a strain of HIV with a defect in a gene called "negative factor" (*nef*). The normal protein product of this gene appears to play a role in regulating the rate of viral replication. The defective gene produces a defective protein that down-regulates viral reproduction to the point that it stops or at least proceeds so slowly that the immune system can produce enough antibodies to keep the virus at bay. What is exciting about discovery of the *nef* mutant is that it offers a possible way of producing a vaccine. Because the *nef* strain appears to have the usual protein structure of the HIV viral coat, it might be used to stimulate the immune system to produce antibodies. Because its reproductive rate is so low, the *nef* itself is not able to infect the body.

Conclusion

We began this chapter by focusing on how to formulate hypotheses and test them either by observation or experimentation. Testing hypotheses by observation is characteristic of many sciences, for example astronomy and paleontology, in which the investigator cannot intervene in the phenomenon under consideration. Although the requisite observations are not always readily available (a specific fossil, for example), new observations are critical to the testing of many hypotheses. Where possible, however, experimentation is preferable. Experiments allow the investigator to intervene in the phenomenon under consideration in a direct and controlled way. We discussed the process of setting up control and experimental groups that were matched in all respects except for the factor being investigated (for example, the role of light in growth of plant seedlings). We also pointed out the advantage of formulating hypotheses in the null form (for example, "*If* light has *no* effect on seedling growth, *then*…), because rejection of such a hypothesis yields a more certain conclusion than does merely supporting it.

We then examined the process of formulating and testing hypotheses in two real-life attempts to understand the cause of human disease: cholera in the nineteenth century and AIDS in the late twentieth century. In both cases, direct experimentation with human subjects to study mechanisms of transmission was not possible. To follow the spread of the disease, researchers in both cases had to work with the "experiments" that nature or social circumstances presented. In the cases presented here in which human subjects were involved, procedural issues, such as comparing populations matched for age, sex, and socioeconomic status, were important aspects of setting up control groups. Ethical issues, too, are critical matters when studying human disease and testing possible cures.

Exercises for Critical Thinking

3.1. Explain why it is always important to formulate hypotheses that can be tested, as opposed to those that, for whatever reason, cannot be tested.

3.2. Give an example of one or more hypotheses that could be tested primarily (or only) by observation. Why are hypotheses that can be tested by experimentation to be preferred over those testable only by observation?

3.3. A number of years ago a study was carried out attempting to correlate body shape in adolescent males and a tendency toward criminal behavior. The researchers assembled two study populations: boys in a Massachusetts reformatory (average age, 16.5 years) and boys in a fashionable public school in Boston, Boston Latin School (average age, 15.5 years). The hypothesis being tested was that a more muscular (mesomorphic was the technical term) body form was associated with criminality, whereas both thin, wiry (ectomorphic) and fat (endomorphic) body shapes were not. The researchers found that their prediction was borne out: The body form of the reformatory population was on average more mesomorphic than the population in Boston Latin School.

    a. In terms of experimental design, what problems do you see with this research plan?

    b. What ethical concerns might you raise about this sort of investigation?

3.4. Discuss why the solution to the spread of epidemic disease is never simply a matter of the technicalities of biology or medicine but also involves social dimensions. Use the cases of cholera or AIDS to illustrate your points.

3.5. Using your answers to question 3.4 as a starting point, discuss the issues of how to balance individual rights versus the collective good in a community (city, state, nation) where an epidemic is in progress. Should people with a contagious disease be quarantined? This issue has come up in some communities where individuals have called for the quarantine of patients with AIDS. Should drugs be legalized to reduce the number of intravenous users who get their injections in back-alley settings in which there is no control over sanitation? Should those entering a community (for example, at immigration ports) be required to have medical tests for communicable diseases before being allowed entry? At the height of a cholera epidemic (known to have come from Europe) in the Mississippi Valley in the late 1840s, steamships planning to dock in St. Louis were required to stop a number of miles below the city and be searched for cholera patients, who were then prevented from disembarking. How ethical or necessary were such restrictions?

Further Reading

On the process of science in general
Goldstein, Martin, and Inge Goldstein. *How Do We Know?* New York, NY: Plenum Press; 1978.
> *Contains a number of case studies from many areas of science, illustrating various aspects of how we know what we know. One chapter focuses on Snow's investigations on cholera.*

On John Snow's study of cholera
Rosenberg, Charles. *King Cholera.* London, UK: Hamish Hamilton; 1966.
> *A series of very well-written essays about the cholera epidemics in London in the nineteenth century.*

On AIDS and transmission of HIV
Duesberg, Peter. *Inventing the AIDS virus.* Washington, D.C.: Regnery, 1996.

Duesberg, Peter. *Infectious AIDS: Have We Been Misled?* Berkeley, CA: North Atlantic books, 1995.
> *These two books present Duesberg's unorthodox hypothesis that AIDS is not directly the result of infection by HIV but by an unhealthy lifestyle that leads to a breakdown in body defenses (allowing infection by HIV virus and other opportunistic pathogens. See also Website 6.*

Hooper, Edward. *The River: A Journey to the Source of HIV and AIDS.* Boston, MA: Little, Brown and Co.; 1999.
> *A recent study dealing with the history of the AIDS epidemic and with speculations on its possible origins. The author leans toward the infected polio vaccine theory but presents fairly the limitations of this hypothesis.*

Montagnier, Luc. *Virus.* New York, NY: W.W. Norton; 1999.
> *A more biologically oriented discussion of the origin of AIDS than the Hooper book. Montagnier's book contains detailed discussions about how the virus itself infects T cells and the various drug strategies that have been employed to try and stop it.*

Suggested Websites

1. HyperStat Online Textbook

http://davidmlane.com/hyperstat/

> *This is a very extensive on-line textbook on statistical analysis and hypothesis testing. While not for the faint of heart, each subtopic provides a variety of very general links to sites that explore statistical analysis in a more fundamental manner.*

2. John Snow

http://www.ph.ucla.edu/epi/snow.html

> *"Devoted to the life and times of Dr. John Snow." this site includes biographical information about John Snow's life, details concerning his discovery of the causes of the 1849 London cholera epidemic, and descriptions of competing notions of cholera etiology.*

3. Centers for Disease Control and Prevention, National Center for HIV, STD and TB Prevention

http://www.cdc.gov/hiv/dhap.htm

> *A huge number of links to internal documents at the CDC concerns all aspects of HIV infection, prevention, therapy, and the progression of the infection to AIDS.*

4. Origins of HIV and the AIDS Epidemic

http://www.uow.edu.au/arts/sts/bmartin/dissent/documents/AIDS/rs/

> *Papers, press releases, media stories and responses to a September, 2000 meeting on the origin of HIV sponsored by the Royal Society of London. Pointed responses to the hypothesis that HIV arose from polio vaccine (Edward Hooper,* The River*) took center stage at this conference.*

5. Duesberg on AIDS

http://www.duesberg.com/index.html

> *Peter H. Duesberg, Ph.D. is a professor of Molecular and Cell Biology at the University of California, Berkeley. He is the leading proponent of the idea that HIV infection is only incidental to the development of AIDS.*

6. The Evidence that HIV Causes AIDS

http://www.thebody.com/niaid/hivcausesaids.html

> *Part of an extensive website devoted to HIV and AIDS (www.thebody.com) this page reproduces a detailed refutation by the National Institute of Allergy and Infectious Diseases of claims that HIV is not the cause of AIDS.*

4

Doing Biology: Three Case Studies

The various procedures for asking questions, framing hypotheses, designing experiments, and analyzing their results discussed in the first three chapters of this book are all components of the process of doing science. Many are involved in all scientific research work, whereas others are used only in special circumstances. In this chapter, we turn to three specific case histories in which these processes of science are brought into play in solving specific problems.

In carrying out scientific research, it is critical to choose the best methods for answering any given question, because a large portion of science involves good judgment about the most practical and most fruitful approach. Using an organism that is difficult to work with or for which adequate controls are impossible to maintain may produce ambiguous if not meaningless results. In the 1930s and 1940s, for example, scientists attempted to use the fruit fly *Drosophila* to study the mechanism by which genes exert their various effects during embryonic development. This effort proved largely unsuccessful, because the insects were too small for the dissections and transplantation of embryonic parts necessary for work on complex, multicellular organisms. Some fascinating work did result, but progress was slow and provided answers for only a limited repertory of questions. Microorganisms such as bacteria and yeast later proved much more effective for investigating more precisely the nature of gene action.

In this chapter, we examine three rather different types of research projects, each requiring a different assemblage of methods to carry out the investigation. Each of these case studies comes from ongoing areas of biological research, and, despite the differing methods each employs, all show an underlying similarity in logic and design.

Formulating and Testing Hypotheses in the Laboratory: The Discovery of Nerve Growth Factor

In the developing vertebrate embryo nervous system, nerve cells (neurons) develop from undifferentiated nerve cells (neuroblasts). These first appear in the region of the embryo that will become the central nervous system, composed of the brain and spinal cord. As neuroblasts differentiate, the axon appears, a long extension ultimately connecting the central nervous system with the peripheral nervous system, those nerve cells that innervate specific tissues, such as muscles or secretory glands. In the twentieth century, two basic questions emerged about the way the nervous system develops. First, how does the axon form? Second, once formed, how do the axons from developing neurons find their way so precisely to their target tissues? These two questions obviously are interrelated; the answer to the first provides clues to the second.

The development of the central and peripheral nervous systems was studied in 1905 and 1906 by Yale University embryologist Ross G. Harrison (1879–1959). In the nineteenth century, three hypotheses had been put forward to account for the origin of the axon. One of these hypotheses maintained that the growing axon was formed by Schwann cells, a chain of cells that produce the myelin sheath providing insulation around the axon. The location of the Schwann cells would therefore determine the location of the future neuron. A second hypothesis proposed that the growing nerve fiber was formed along a preexisting "protoplasmic bridge" that acted as a kind of roadway or guideline. A third hypothesis suggested that each nerve fiber was an outgrowth of a single neuroblast, extending outward without the aid of either Schwann cells or protoplasmic bridges.

In 1905, Harrison set out to test these three hypotheses. He needed to find a way to observe neuroblast development outside the embryo's environment, where Schwann cells or the postulated protoplasmic bridges could exert their influences. Harrison isolated neuroblasts from frog embryos and placed them in a drop of sterilized frog lymph fluid, which served as a culture medium. He then suspended this drop from the underside of a deep-well slide (Fig. 4.1), which he could examine under the microscope.

When first placed in the hanging drop, the neuroblasts had not yet begun to differentiate. If the first and second hypotheses were correct, the neuroblast should show relatively little differentiation in terms of axonal outgrowth, because neither Schwann cells (hypothesis 1) nor protoplasmic bridges (hypothesis 2) were present. Only if hypothesis 3 were correct would normal development of a recognizable axon be predicted. The deductive logic of Harrison's experiment may be stated as follows:

If . . .   neuroblasts do not require Schwann cells or protoplasmic bridges to form axons, and

if . . .   neuroblasts are grown in a culture medium in which Schwann cells and protoplasmic bridges do not exist,

then . . . neuroblasts should develop normal axons.

Harrison observed his neuroblast cultures for periods of up to 9 hours. The results of one preparation, observed at 2:50 PM, 4:45 PM, and 9:15 PM on the same day are shown in Figure 4.2. Note that axonal development occurred as a distinct outgrowth of the neuroblast cell body. Clearly, neurons possess their own program for growth and differentiation of the axon. The results contradicted hypotheses 1 and 2 and supported hypothesis 3.

Harrison's work, which was nominated once and considered twice for the Nobel prize, was technically and conceptually brilliant. (The Nobel

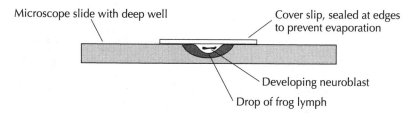

Figure 4.1 Ross Harrison's culture preparation for observing growth and differentiation of an isolated neuroblast (1907). A deep-well microscope slide provided a miniature culture chamber, on the top of which was placed a cover slip that could be sealed at the edges to prevent drying out. The undifferentiated neuroblast was placed in a drop of sterile frog lymph and observed every few hours for a period of 8–10 hours.

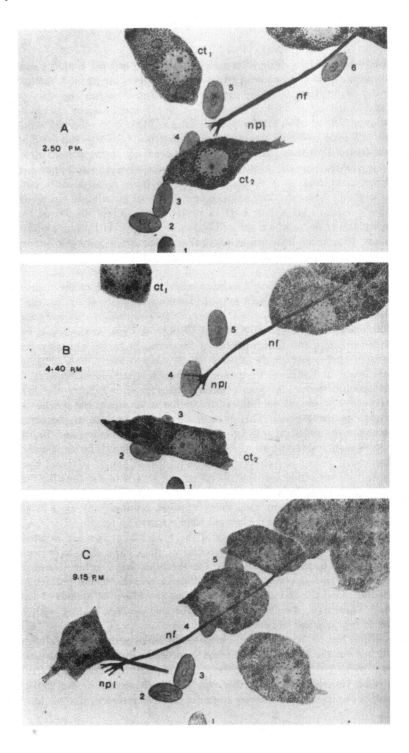

◀ **Figure. 4.2** Development of a living nerve fiber (the axon of the neuron or nerve cell) as observed by Harrison in 1910. (A) Axon growth is depicted at 2:50 PM; (B) 4:40 PM; and (C) 9:15 PM on the same day. The observations clearly show the outgrowth of the axon from the neuroblast, with no help from guiding "protoplasmic bridges" or Schwann cells. The other cells shown are scraps that were transplanted along with the neuroblast from the frog egg.

committee decided against awarding Harrison the prize because he did not follow up on the technique or use it again outside this preliminary set of experiments.) It laid the foundation for techniques of culturing living nerve cells in the laboratory. However, as often happens in science, Harrison's results now raised a second question: How do the outgrowing axons find their way to their appropriate target sites?

One of the first insights into this question came in 1909 from the work of M. L. Shorey. He found that removal of a chick limb bud, that portion of the embryo destined to become a leg or wing, caused a failure of the neuroblasts that would normally innervate the bud to develop. For the next 25 years, nothing much was made of this observation. Then, in 1933, Viktor Hamburger, a German emigré working in the laboratory of F.R. Lillie (1870–1947) at the University of Chicago, approached the same problem in the reverse direction. Instead of removing the limb bud, he implanted a limb bud from one chick embryo into the limb bud region of another, producing an embryo with two limb buds in the same place. The result was that the second embryo produced far more nerve growth extending into the limb region than occurred in the control, an embryo with only one limb. Both Shorey's and Hamburger's findings suggested that nerve growth was somehow regulated by stimuli from the peripheral target zones, such as a developing limb bud. The precise nature of this stimulus remained a mystery. However, in concluding his 1934 paper on the chick experiments, Hamburger put forth a model with several associated hypotheses that suggested how this influence might occur. He proposed that: (1) The developing peripheral regions produced some sort of specific chemical that stimulated neuroblasts to grow in the direction of the developing tissues; (2) The effect must be transmitted from the peripheral region by way of the outgrowing ends of the neurons back up the axon to the cell body, where growth was controlled; and (3) The stimulating effect of the limb bud tissue was quantitative, that is, the amount of stimulating effect was directly proportional to the quantity of limb bud tissue present.

After reading Hamburger's paper, a young medical researcher in Italy, Rita Levi-Montalcini, decided to determine exactly how removal of the target tissues affected the growth of developing neuroblasts. Although she had received a medical degree, Levi-Montalcini could not obtain a university or government position in Benito Mussolini's fascist Italy, because she was Jewish. She performed her experiments in her own home, with fertile chicken eggs supplied by a local dealer. As a result of her observations, Levi-Montalcini argued that, if the limb buds or other target areas of a developing embryo are removed before neurons enter the region, large numbers of the neurons developing in that area actually degenerate. Levi-Montalcini suggested that the peripheral target tissues might function more to *maintain* neuronal growth than to stimulate it. Published in 1942 and 1944, Levi-Montalcini's papers came to Viktor Hamburger's attention. In 1946 Hamburger, who by that time had moved to Washington University in St. Louis, Missouri, invited Levi-Montalcini to join him in his laboratory. In their first experiments together, Hamburger and Levi-Montalcini confirmed that removal of the target areas did indeed cause an increasing deterioration and death of developing neurons. This confirmation supported Shorey's and Hamburger's earlier suggestion that the target tissues must provide some sort of signal for neural development, direction, and survival.

What was the nature of that signal? Some insight into that question came from an unexpected source. One of Hamburger's former doctoral students, Elmer Bueker, then at Georgetown University in Washington, DC, had removed limb buds from chick embryos and replaced them with mouse tumors. The purpose of the experiment was to determine whether a homogeneous tissue such as a tumor would produce the same or different effects on neuronal growth from those produced by the more complex differentiating tissues of a limb bud. Bueker found that one of his tumors, mouse sarcoma 180, grew quite well in chick embryos and that adjacent developing nerves quickly invaded the tumor mass, causing the nerve cell bodies to increase in number by about 30%. Learning of Bueker's work, Hamburger and Levi-Montalcini secured his permission to use mouse sarcoma 180 to investigate how peripheral tissue might provide a specific signal for neuronal growth. They found first that the mouse sarcoma was far more effective than limb bud tissue in stimulating neuronal growth in embryos into which it had been transplanted. They also found that the effect was highly specific: The sarcoma affected primarily sensory neurons. Finally, they found that this effect could be produced even when the tumor was placed outside the major membranous coverings of the egg, through which no cells or large structures could pass. The latter result suggested clearly that the tumor must be producing some substance that diffused through the membrane and affected neuronal growth.

To identify what that substance might be, Hamburger hired a post-doctoral fellow, Stanley Cohen, then at the biochemistry department at Washington University School of Medicine, to work with Levi-Montalcini. First, the investigators prepared cell-free extracts from the tumor. They found this extract stimulated neuron growth in exactly the same fashion as the intact tumor and that it contained a mixture of proteins, nucleic acids, and other macromolecules. The active component, whatever it might be, was named "nerve growth factor" (NGF).

To determine the specific nature of NGF, Cohen and Levi-Montalcini began a process of eliminating the various possible components of the active mixture. They first treated their extract with snake venom, which contains an enzyme that degrades nucleic acids, reasoning that, if the extract lost its ability to stimulate neuron growth after treatment, then NGF must be a form of nucleic acid. These experiments involved growing nerve cells in culture medium dishes with NGF, NGF and venom, and venom alone (Fig. 4.3). To their complete surprise, Cohen and Levi-Montalcini found that neural growth was greater in the control (that is, venom alone) than in the culture medium containing venom plus tumor extract. It was clear, therefore, that NGF was not a nucleic acid. In fact, it turned out that snake venom itself is a rich source of NGF. The results also supported Levi-Montalcini's hypothesis that NGF serves primarily a maintenance role in neuronal development. Many neurons begin to grow out from the central nervous system. Most, however, degenerate before reaching a target site. The difference between which ones degenerate and which ones continue to grow has to do with whether or not they come into contact with NGF early in their differentiation. Additional studies identified NGF in mammalian salivary glands, the anatomical counterpart of snake venom glands. NGF could be found especially easily in the salivary glands of male mice.

Armed with these rich sources of NGF, Cohen and Levi-Montalcini began an all-out assault on its precise identity. Hamburger, meanwhile, returned to his main interest, factors determining nerve growth patterns during chick development. During the mid and late 1950s, Cohen and Levi-Montalcini showed that NGF was a protein, and its amino acid sequence eventually was determined. More recently, the DNA for NGF has been sequenced and cloned for a number of species. Before leaving St. Louis for Vanderbilt University (Nashville, TN) in 1960, Cohen discovered yet another growth factor, known as epidermal growth factor (EGF). This discovery, along with NGF, suggested that vertebrate development might be guided by a series of growth factors emanating from various peripheral tissues.

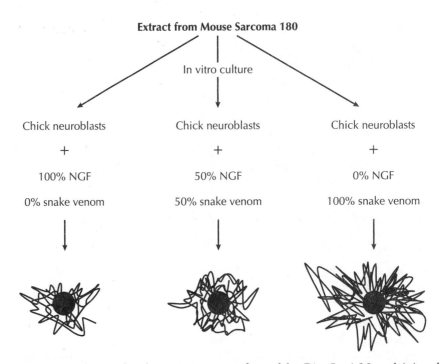

Figure 4.3 Design for the experiment performed by Rita Levi-Montalcini and Stanley Cohen to demonstrate the effects of NGF on chick neuroblast proliferation. An extract of mouse sarcoma 180 was prepared containing NGF. This was then administered by itself to chick neuroblasts in cell culture (left, 100% sarcoma NGF) to chick neuroblasts in combination with snake venom (middle, 50% mouse sarcoma NGF and 50% venom), and with no mouse sarcoma NGF and only snake venom (right, 0% NGF and 100% venom). The experiment showed clearly not only that NGF had an effect on proliferation, but also that snake venom had an even more powerful effect, a surprising outcome. In this set-up, each experiment served as a control for the others.

Although the exact mechanism of NGF action is still under investigation, it is clear that neurons do not have to find their way to target tissues unaided. NGF plays an important role in enabling them to get there. For their work on NGF and its extension to EGF and to other growth factors, Levi-Montalcini and Cohen were awarded the 1986 Nobel prize in physiology and medicine.

Research in the Field: Homing in Salmon

The investigation of NGF is a classic case of laboratory-based research in the biological sciences. The experiments were performed under conditions allowing for the control of all variables that might have affected the oucomes.

At times, however, experiments must be carried out under conditions in which it simply is not possible to control all the variables that might influence the outcome of a research study. Such experiments are often referred to as being carried out "in the field." Here, we will analyze one such field research project.

The Organism

Few fishes are better known than the salmon. Its characteristic pink meat has been enjoyed by untold millions of humans. Archeological discoveries from human riverside habitats dating back thousands of years suggest that this enjoyment is very old. The Atlantic salmon (*Salmo salar*) can be found in the waters of virtually every European country bordering the Atlantic ocean and in those of countries forming a great arc westward across the Atlantic (Iceland, Greenland, Newfoundland, Canada) to the northeastern portion of the United States (Maine). The silver salmon (*Oncorhyncus kisutch*) is prized by both sport and commercial interest groups and is found in waters extending from the northwest corner of the United States into western Canada and Alaska.

Beside the pink color of their meat, both Atlantic salmon and Pacific salmon share certain other features. One of these shared characteristics is an unfortunate one: a drastic decline in population (Fig. 4.4). Exactly *why* this is the case remains unclear, and numerous hypotheses have been put forth: overfishing, rising water temperatures as a result of global warming,

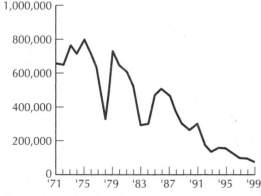

Source: Atlantic Salmon Federation

Figure 4.4 Decline in the population of the Atlantic salmon (*Salmo salar*) from 1971–1999. Similar population declines have been documented for its western relative, the silver salmon (*Oncorhyncus kisutch*). The precise cause or causes of this decline remain uncertain.

disease and parasites introduced by escaped farm-raised fish (whose caged conditions tend to encourage disease and parasites), and increased salmon kills by seals (whose numbers have increased through the effects of protective laws).

A second feature shared by both Atlantic and Pacific salmon is that they are both migratory species. Atlantic salmon hatch in freshwater streams, and, after reaching a certain size and in response to some as yet undetermined stimulus, the fish swim downstream to the Atlantic ocean for several years of feeding and growth (from 5 to more than 20 pounds). Upon reaching sexual maturity, they then return to freshwater streams, where females lay their eggs (spawn) to be fertilized by males. After hatching, the cycle begins again. Similarly, silver salmon hatch in the freshwater streams of the Pacific Northwest. The young fish swim downstream to the Pacific Ocean, where they may spend 5 or more years attaining full size and sexual maturity. They then return to freshwater streams, often jumping incredible heights up waterfalls to reach the spot at which they lay their eggs (Fig. 4.5). As in the case of the Atlantic salmon, the females lay their eggs, and the males fertilize them. The adult fish then die.

By tagging 2- and 3-year-old silver salmon at the stage (smelt) when they are ready to go from their home streams to the ocean to mature, biologists have found that, to begin the life cycle over again, salmon return to the precise stream from which they hatched. Note that this last observation is one based upon a finite number of observations, that is, a sampling of the silver salmon population studied, and is therefore merely an inductive generalization (see Chapter 2). Assuming this generalization is correct, a deeper question now arises: *How* do the salmon find their way to the stream where they hatched? In other words, what is the causative factor involved in leading adult silver salmon to return precisely to their home streams instead of to any other freshwater stream?

Two Possible Hypotheses

Two possible explanatory hypotheses might account for the ability of salmon to locate their home streams:

Hypothesis 1: Salmon find their way back by the sense of sight; that is, by recognizing certain objects they saw when they passed downstream on their way to the sea.

Hypothesis 2: Salmon find their way back by their sense of smell, that is, by detecting certain substances and or proportions of those substances that characterize their home streams.

Figure 4.5 Silver salmon leaping up a waterfall as they swim up the freshwater streams where they will reproduce. In the case of the Alaskan sockeye salmon, it has been estimated that only about 1 fish out of 1000 make it successfully to the spawning grounds. [© B. & C. Alexander/Photo Researchers.]

Several other hypotheses also are possible, of course. Indeed, it may well be that both hypotheses are correct and that salmon use their sense of sight *and* smell in locating their home stream. However, it is more practical at least to begin by working with just these two alternative hypotheses and designing experiments to test them. For example, if hypothesis 1 is correct, then salmon with shields placed over their eyes should be unable to find their way home. This reasoning may be expressed deductively as follows:

If . . .   *Oncorhyncus kisutch* salmon use visual stimuli alone to find their way to their home streams to spawn,

then . . .   blindfolded salmon of this species should not be able to find their way home.

In fact, blindfolding experiments reveal that the fish find their way home just as well as they did before. If we assume that no variables that might have influenced the results were overlooked, have these blindfolding experimental results *disproved* our hypothesis? Yes. Suppose, on the other hand, that the blindfolded fish did *not* find their way to their home streams. Would these results *prove* the hypothesis? No. As we stressed in Chapter 2, it lies beyond science to "prove" anything. The experimental results can be said only to *support* hypothesis 1 or, at the very least, to be consistent with it. Although this would support the visual hypothesis, such results would not rule out the role of smell.

Now let us test hypothesis 2:

If . . .   *Oncorhyncus kisutch* salmon find their way to their home stream by following its distinctive chemical characteristics upstream,

Then . . .   blocking the olfactory sacs with which fish detect odors or tastes should prevent salmon of this species from locating their home streams.

This experiment was performed several years ago by biologist A. D. Hasler and his associates of the University of Wisconsin, working with silver salmon from the Issaquah Creek and its East Fork, two streams near Seattle, WA (Fig. 4.6). Note that these streams join together just before they empty into Lake Sammanish. A total of 302 salmon were captured for the experiment. Approximately equal numbers came from each stream. The fish were then divided into four groups. Two control groups were established, one from each stream. Each fish was tagged to indicate the stream from which it came. The experimental groups were also tagged. In addition, their olfactory sense was disrupted by the insertion of cotton plugs coated with petroleum jelly or the application of benzocaine ointment (an anesthetic) into their olfactory pits (Fig. 4.7). The sensory nerve endings responsible for their sense of smell are located in the olfactory pits. These experimental group fish were released .75 mile downstream from the juncture of Issaquah Creek and the East Fork and then recaptured at traps set about 1 mile above the junction.

Tabulating the Results

As Table 4.1 shows, not all the released fish were recaptured upstream. Some swam downstream after release, whereas others swam upstream but missed the traps. Only 50% of the controls and 45% of the experimental fish were recaptured. More important, however, were the results of the distribution of fish according to their stream of origin. Table 4.2 indicates a

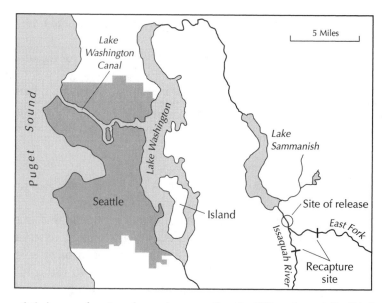

Figure 4.6 A map showing the region near Seattle, WA, where A.D. Hasler and his research team conducted field studies on factors influencing homing behavior in the silver salmon.

significant difference between the control and experimental groups in terms of finding their way home. Among the controls, 100% of the Issaquah Creek and 71% of the East Fork salmon found their way back to their home streams. In the experimental groups, however, only 77% of the Issaquah Creek and 16% of the East Fork salmon found their streams of origin. Thus

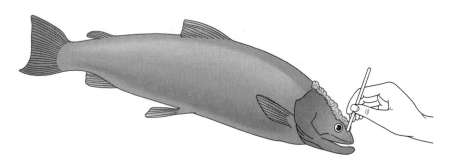

Figure 4.7 Inserting a cotton wad (visible above the eye of the fish) permeated with petroleum jelly or benzocaine ointment into the olfactory pit of an anesthetized and tagged silver salmon.

Table 4.1 Total silver salmon tagged compared to number captured from experimental and control groups.

	Total tagged	Recaptured		Not captured	
		Actual number	Percent	Actual number	Percent
Control	149	73	49.6	76	50.4
Experimental	153	70	45.0	83	55.0

hypothesis 2, maintaining that the olfactory sense is responsible for the ability of salmon to identify their home streams, was supported.

Note that *some* fish from each experimental group *did* find their way back to their home streams, a result that might appear to contradict hypothesis 2. Here, however, the laws of chance and probability come into play. A certain number of the fish would be expected to return to their home streams purely by chance. The number of experimental salmon that found their way to the home stream was not significantly greater than the number that could be expected to get there purely by chance. Thus, hypothesis 2 is still valid.

Additional Questions

The salmon experiment demonstrates a characteristic of any scientific problem-solving activity: The answer to one question often raises many more. Hasler's experimental support for hypothesis 2 immediately raises

Table 4.2 Distribution of recaptured silver salmon, comparing experimental and control groups.

	Recapture site			
	Issaquah		East Fork	
Capture site	Actual number	Percent	Actual number	Percent
Controls				
Issaquah (46 fish)	46	100	0	0
East Fork (27 fish)	8	29	19	71
Experimentals				
Issaquah (51 fish)	39	77	12	23
East Fork (19 fish)	16	84	3	16

Data from A. D. Hasler, *Underwater Guideposts*, Madison: University of Wisconsin Press, 1966, p. 42.

another question: What is it in their home streams that the salmon detect by their sense of smell? What sort of factor or factors in the stream serve as a guide? A number of hypotheses come to mind:

Hypothesis 1: Silver salmon detect discrete combinations of dissolved inorganic substances (minerals) in the water of their home streams.

Hypothesis 2: Silver salmon detect discrete combinations of dissolved organic compounds characteristic of their home streams.

Hypothesis 3: Salmon somehow are able to recognize other salmon of their own home stream population.

Again, using deduction, each of these hypotheses leads to specific predictions that, in turn, may be verified or refuted experimentally.

Studies with other closely related migratory organisms also may help. For example, it has been shown that eels, which also migrate from freshwater to saltwater (where they spawn) and return, are enormously sensitive to dissolved minerals and organic material in water. A single eel can detect the presence of a substance dissolved in water and diluted to the extent that only two or three molecules per liter of the substance are present (Fig. 4.8), an observation lending support to hypotheses 1 and 2 in the case of the silver salmon.

Fortunately, some experimental data are available to help distinguish among hypotheses 1, 2, and 3. Dr. D.J. Solomon of the Ministry of Agriculture, Fisheries, and Food in England studied Atlantic salmon migration around the Usk, Wye, and Severn rivers. These rivers empty into the Bristol Channel on the west coast of England (Fig. 4.9). The adult salmon breed and the young fish mature in the upstream portions of each of these rivers. Several other rivers also empty into the Bristol Channel but do not contain salmon populations. Solomon became curious about what attracted adult salmon into the Bristol Channel, and thence into the Usk, Wye, and Severn, but kept them from invading other available rivers. He reasoned that if something as general as dissolved minerals and organic matter were the attractant, then one would expect to find Bristol Channel salmon swimming selectively into the point of entrance of freshwater from *all* the rivers emptying into the channel's saltwater estuaries. Because ocean water contains far more dissolved minerals and organic matter of the same general type found in freshwater rivers, fish would have to swim part of the way up any river to be able to determine whether it was "home." Yet, in a survey of 23,000 tagged fish, only six were found swimming into those

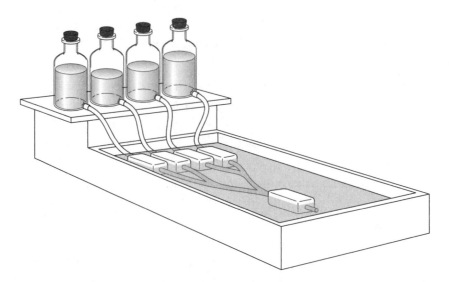

Figure 4.8 Whenever possible, biologists attempt to construct an apparatus that will allow testing hypotheses in the laboratory rather than in the field. By so doing, variables that might influence the results are more easily controlled. This apparatus tests the ability of eels, another migratory fish, to detect minute amounts of dissolved substances in water. The object is to see whether these substances play a role in young eels' ability to find their way from the Sargasso Sea, the Atlantic Ocean region where they hatched, to the freshwater streams where they spend most of their lives. It was shown that young eels show no preference for tap water over sea water but had a strong preference for natural inland water to sea water, being able to detect certain organic compounds in the water even when diluted to 3×10^{-29} parts per million. This means they must be reacting to the presence of only two or three molecules of the substance. Because eels are a different species of fish than salmon, such results provide only indirect support for the olfactory hypothesis about how silver salmon are able to locate the stream in which they hatched.

rivers without salmon. It appeared that the salmon were able to detect their home rivers long before they entered the mouths of those rivers, and, therefore, some other factor or factors must be at work. This observation led Solomon to formulate another hypothesis: The main factor attracting salmon to a river is the presence of other salmon. He knew that many animal species are known to produce pheromones, distinctive substances by which members of each species recognize one another. Solomon hypothesized that the factor attracting salmon to one salmon-containing river over another is the pheromone produced by that river's specific population of young salmon.

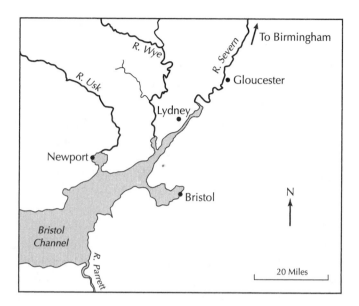

Figure 4.9 A map of the west coast of England, showing the Bristol Channel with the rivers Usk, Wye, and Severn emptying into it. These three rivers normally contain salmon, but other rivers also emptying into the Bristol Channel generally do not.

Solomon needed a way to test this hypothesis. Designing an experiment in nature with such a huge population would be difficult and involve innumerable practical obstacles, not the least of which would be expense. However, in going through the research literature, Solomon discovered data from a 20-year-old breeding experiment that provided a test for his hypothesis. In the 1950s, the British government had financed an attempt to establish a salmon breeding ground in a tributary of the Parrett River, which also empties into the Bristol Channel. The attempt ultimately met with failure when no salmon population became established in the Parrett. However, records showed that, in the 3 years immediately after the introduction of the first breeding population, the number of adult salmon found swimming up the Parrett increased several fold. Most significant, these fish had not come from the newly established fishery population but rather from the Usk, Wye, and Severn populations. These fish appeared suddenly to have found the Parrett interesting! Using deduction, in this case retroactively, Solomon could see that a prediction based upon his pheromone hypothesis had already been verified by the Parrett River study 20 years before. Expressed hypothetico-deductively:

> *If . . .*   Atlantic salmon recognize pheromones produced by
> other salmon,
>
> ***then*** *. . .*   introducing young salmon into a river ordinarily without
> salmon should increase the chances of adult salmon
> swimming into that river.

As already noted, the Parrett River data supported Solomon's hypothesis. As we know from Chapter 3, however, a true prediction does not necessarily mean a true hypothesis. Nonetheless, the data *do* lend support to the hypothesis by verifying one of the predictions that follow from its acceptance.

Testing the Pheromone Hypothesis Further

It had been shown as long ago as 1938 that salmon transplanted at the premigratory stage from their native river to a second river returned as adults to the second river. In a contrasting set of experiments, salmon were kept in a hatchery until well past the age at which they would have begun their journey from their home streams to the ocean. When released into different streams, they failed to return as adults to their transplant stream, returning instead in a random pattern to a number of different streams. These results suggested that a critical period must exist in a young salmon's life during which it learns to recognize the stream in which it hatched. Further experiments, conducted between 1955 and 1971 by a number of investigators, found that young salmon are especially receptive to learning the chemical characteristics of their home stream just at the time they enter the stage of their life at which they begin to migrate toward the ocean. In fact, during this critical period, exposure to these chemical odors for as short as 4 hours is enough for the young salmon to "remember" this odor for life, a phenomenon referred to as imprinting. Thus, it could be hypothesized that, at a critical period in life, young salmon become imprinted with the specific chemical composition of their home streams and later use this imprinted memory to guide them back as adults.

In the mid-1970s, along with Allen Scholz and Ross Harrel, A.D. Hasler decided to study the imprinting process directly. They exposed a group of young salmon at the premigratory stage to morpholine (C_4H_9NO), a nontoxic organic chemical. They then selected two groups of salmon that had been raised in a hatchery until the premigratory stage. They exposed the experimental group to low dosages of morpholine (1×105 mg/L) for a month. The control group received no morpholine. Both groups remained in the hatchery for 10 more months, after which they were brought to the laboratory, placed in tanks with morpholine, and their brain waves recorded. As the research team had predicted, the experimental group that had been exposed to morpholine showed a significantly higher amount of brain wave activity than did the controls that had not been exposed.

Hasler and his co-workers then attempted to resolve another question: Will fish exposed to morpholine actually swim toward streams into which the compound has been placed? They took the experimental and control groups of salmon described above and released them into Lake Michigan at a point equidistant between the outlets of two creeks. Into Owl Creek they placed a concentration of morpholine about equal to that maintained in the hatching tanks. The other stream was left undisturbed. The fish were marked, and 18 months later the investigators carried out recapturing procedures to see which fish were found in each stream. Out of a starting population of 16,000, a total of 246 tagged fish were recaptured in Owl Creek, about the expected percentage of recapture in other such experiments. Of the 246 fish recaptured, 218 had been exposed to morpholine, whereas only 18 were from the control group. In the other creek, most of the fish were from the control group. It seemed clear that the cue that had attracted such a disproportionate number of the experimental group to Owl Creek was the presence of morpholine in the water.

Could something else in Owl Creek have served as a cue to the experimental fish? In a second experiment, representatives of the control and experimental groups were released near the two streams, but no morpholine was added to either stream. If the experimental fish had found their way into Owl Creek because of the presence of some other cues, they ought to do so in the same approximate numbers in this experiment as in the first. Hasler and his co-workers found about equal numbers of both control and experimental fish in each creek. Therefore, the hypothesis proposing that salmon became imprinted to a specific chemical compound (or group of compounds) early in life seemed well supported.

In yet another experiment, Hasler and his associates attempted to track the movement of adult salmon through the water in response to the presence of morpholine. They implanted ultrasonic transmitters as tracking devices into 40 salmon, 20 of which had been imprinted on morpholine and 20 of which had not. After releasing the fish into Lake Michigan, the researchers set up an "odor barrier" by pouring a line of morpholine in the water along the shore. It was found that the fish exposed to morpholine would stop swimming when they encountered the odor barrier and mill around within the barrier for up to 4 hours, until the scent had been washed away by the current. The control salmon, not exposed to morpholine, swam through the barrier without stopping.

From all of these results, Hasler felt that he could draw three major conclusions:

1. Salmon use their olfactory sense to detect the presence of familiar substances in water in identifying their home streams.
2. Salmon become imprinted to the familiar chemical odors of their home streams when in their early premigratory stage.
3. The specific chemical odors to which young salmon become imprinted guide them as adults in returning to their home streams.

"Dissecting" the Experiments

The work of Hasler and Solomon with salmon illustrates several important features of testing hypotheses in science. Note that both investigators framed their questions in ways that were testable. They did not put their question in a general form, such as, "How do salmon find their way home?" Instead, they broke this large question down into several smaller, more directly answerable questions, among them, "Do salmon use their olfactory sense to find their way home?" and "Will salmon become imprinted to an artificial substance if they are exposed to it early in life?" Both questions suggest observations or experiments that might provide answers.

A second point about scientific research illustrated by this case is the importance of approaching a given problem from a number of directions. Hasler and his research team were not content simply to perform one or two experiments testing the olfactory hypothesis. They showed not only that salmon use their olfactory sense to find their way home as adults but also that salmon become imprinted to specific substances early in life, show a distinct neurological reaction to substances to which they have become imprinted, and select streams containing these substances when they return. All these lines of evidence converge to support the initial hypothesis 2, which proposed that young salmon use their olfactory sense to "recognize" the chemical peculiarities of their home streams and use this recognition to guide them back to these streams as adults.

A final point concerning scientific methodology illustrated by the salmon experiments comes from Solomon's work. It is not always possible in science to perform an experiment to test a hypothesis. Yet it often happens that an observation or group of observations will serve the same purpose. Solomon's observation that adult salmon appeared in the Parrett River only after young fish had been introduced there supported one aspect of the olfactory hypothesis, in this case, the idea that part of the olfactory stimuli to which adult salmon react may be pheromones or some similar substance or substances resulting from the presence of other salmon. Solomon did not perform an experiment to come to this conclusion. The experiment, in a way, had already been done for him. He simply took advantage of already-collected data that were relevant to the question he was asking.

Statistical Significance

Quantitative data are a cornerstone of modern science (see Appendix 1). Yet it is obvious that a large collection of quantitative data is of little or no value if it is not arranged in such a way as to demonstrate important relationships. The individual data for each tagged fish in Hasler's experiments yield little information if they are presented randomly. In collecting data in the field, of course, Hasler and his associates had to tabulate it as it came in. As they captured a fish, they would record where it had originated, whether it belonged to the experimental or control group, and the stream in which it had been caught. Their field notebooks might have contained information arranged as is shown in Table 4.3. Note that these data are both quantitative and qualitative. In the silver salmon experiments, both quantitative and qualitative data were collected. Dr. Hasler's first experiment involved quantitative data. He counted the number of each fish population recaptured in each river and expressed the number in terms of percentages of those released.

Table 4.3 Hypothetical table of raw data as it might have been recorded by A. D. Hasler and his research associates.

Fish number (in order of recapture)	Control group		Experimental group	
	Where released	**Where recaptured**	**Where released**	**Where recaptured**
1	Issaquah	Issaquah		
2	East Fork	Issaquah		
3			Issaquah	East Fork
4	East Fork	East Fork		
5			East Fork	East Fork
6			East Fork	Issaquah
7	Issaquah	Issaquah		
8			Issaquah	East Fork
etc.				

As presented thus far, these data do not allow us to determine readily whether the odor hypothesis (hypothesis 2) has been confirmed or disproved. By arranging the data to allow a comparison between the experimental and control groups, Hasler was able to determine that a difference did exist. Silver salmon with their olfactory sacs plugged found their way back home less frequently than those with their olfactory sense left unimpaired (Table 4. 3). What, then, is the function of the data shown in Table 4. 2? This table records the difference between the number of fish released in each group and the number recaptured. Note that for both groups the percentage of fish recaptured (or conversely, not recaptured) is approximately the same (49.6% of controls, 45% of experimentals). This is vitally important information. If a significant difference in recapture rate between experimentals and controls had been found, comparison of the data shown in Table 4.3 would not be very useful. If a statistically significant decrease in the percentage of silver salmon in the experimental group, compared to the controls, had found their way home, we would have to conclude that the plugging of their olfactory pits must have affected the general ability of the fish to navigate. The similar recapture rate recorded for both experimental and control fish, however, indicates that the two groups must have had an approximately equal ability to navigate.

Table 4.3 also informs us whether both experimental and control groups were subject to the same sampling error. Suppose, for example, that only 10% of the total experimental group had been recaptured as compared with, say, 50% of the controls—obviously a very large difference. Because one of the figures, 10%, is quite small, it might not be representative of the whole. In fact, the number of fish known to have arrived at either their home stream or at another is subject to sampling error, a factor that enters in when we consider only a small number of cases out of a very large population. The larger the sample taken, the less significant the sampling error is likely to be. In Hasler's experiment, sampling error would not be much of a factor if the recapture rates of the two groups were between 80% and 60%. Because Hasler was dealing with a total of only about 150 fish in each group, control and experimental, a large difference in recapture rate between the groups could have led easily to sampling error in the smaller group. Determining that the recapture rates were similar for both groups eliminated this problem.

Recall, however, that *some* of the experimental fish, in fact, *did* end up in streams that were not their home streams. Only in the case of the Issaquah Creek control group did 100% of the fish manage to get back to the stream where they hatched. Among the East Fork controls, only 71% returned to the East Fork. This confirms that the inductive generalization "salmon return to their home stream to spawn" is accurate only in a statistically significant number of cases. Precisely the same applies to most situations in the real world. Although from a strictly logical point of view a single exception to an inductive generalization is enough to make it invalid, in nature this is not the case. The generalization that salmon find their way home by the use of environmental clues obtained through their olfactory sense does not necessarily mean that other factors may not be involved. Predators, water currents, fatigue, or the behavior of other nearby fish also may have affected the direction Hasler's silver salmon took when they came to the juncture of the Issaquah Creek and the East Fork. Thus, although it generally may be correct that silver salmon find their way home by the use of environmental clues obtained through their olfactory sense, this does not mean that other factors do not influence an individual salmon's migratory direction.

Here again is where the statistical significance of the difference enters into the equation. Table 4.1, for example, shows that, of all the fish released, about 50% of the control group and 45% of the experimental group were recaptured. Is this difference statistically significant? Under some circumstances it might be. To Hasler, who knew the conditions under which the fish were tagged, released, and recaptured, such a difference appeared to be insignificant. He was confident that the rates of recapture for the experimental and control groups were virtually the same. A statistical significance test verified this conclusion, but that does not mean that Hasler was absolutely right. Sometimes small, unexpected differences between what is predicted and what is actually found to be the case lead to new discoveries. A highly subjective element in science lies in knowing when to ignore and when to pay attention to such differences.

The same test of statistical significance may be applied to the data in Table 4.2. If the olfactory hypothesis were correct, for example, we might expect that, with very large samples, approximately 50% of the Issaquah Creek group would end up in the Issaquah Creek and 50% in the East Fork, with similar distributions for the East Fork group. With no olfactory sense to guide them, silver salmon in the experimental group should enter streams more or less at random. The data show that, among the experimental groups, more Issaquah Creek fish ended up in the Issaquah Creek than in the East Fork (77% and 23%, respectively), an approximately 25% error between the actual and the expected results. Among the East Fork experimental subgroup, the situation was quite different. Here, 84% ended up in their nonhome stream, the Issaquah Creek, whereas only 16% arrived at their home stream, the East Fork. Are these differences from expectation (50:50 in both cases) statistically significant? The situation is most clear-cut in the East Fork experimental group. Of these fish, 84% ended up in their non-home stream, whereas by chance we would expect about 50% to do so. Comparing the East Fork experimental group with their control group counterparts, we see that in the latter only 29% ended up in the wrong stream. Here, no statistical test is necessary to state conclusively that the difference between 29% and 84% is statistically significant. However, data from the Issaquah Creek experimental and control groups presents an interesting twist. It is certainly an unusual result that 100% of the Issaquah Creek controls were recaptured where they were predicted to be. Among the experimental group, only 77% were recaptured in the home stream, a figure also considerably higher than the 50% we might expect on the basis of chance alone. Indeed, the Issaquah Creek experimental group with no olfactory sense actually did *better* than the East Fork controls with fully functional olfactory sense! Quite obviously, Hasler's conclusion that the data still supported the olfactory hypothesis came to some extent from considerations other than hard data. This example emphasizes the many arbitrary and often subjective elements that may enter into data analysis and the evaluation of scientific hypotheses based on these data.

An Evolutionary Historical Case: Mass Extinctions and the End of the Dinosaurs—The Nemesis Affair

In the late 1970s, geologist Walter Alvarez, at the University of California, Berkeley, became interested in events occurring at major transition points in the earth's history. One such transition was that between the Cretaceous and Tertiary periods, approximately 65 million years ago. The K–T boundary (the letter K is used, rather than C for this scientific shorthand, to avoid confusion with the Cenozoic period) marks the end of the Mesozoic and beginning of the Cenozoic eras, a time known not only for the rapid extinction of the dinosaurs but also of many other forms of life, truly a period of mass extinctions. Alvarez was working on a site in northern Italy, near the town of Gubbio, where the actual boundary layer is exposed. He noted an unusual thin layer of clay between two thicker rock strata. To determine how long it took to deposit this thin clay layer, Alvarez, along with his father, Luis Alvarez, a physicist also at Berkeley, noted and measured the amount of the rare metal iridium (Ir) in the clay. Iridium is found in very small concentrations on earth but is known to be far more plentiful in meteorites and other extraterrestrial objects. Iridium enters the earth's atmosphere at a fairly regular rate in the form of a shower of cosmic dust, and thus its concentration in a layer of sediment would provide a good estimate of sedimentation rate. When the clay was analyzed, it was found to have abnormally high concentrations of iridium: 10 parts per billion (ppb) compared with 3 ppb in the rock strata on either side. Additional information on other K–T boundary site sediments showed iridium to be abnormally high in such sediments around the world (Fig. 4.10).

What could have caused such an unusual and widespread deposition of iridium 65 million years ago? In a 1980 paper in the journal *Science*, the Alvarez team proposed a novel explanation: the high level of iridium was the result of the impact of a giant asteroid, roughly 10 km (6 miles) in diameter, that collided with the earth, stirring up clouds of dust and ash and causing widespread volcanic activity. The resulting atmospheric ash reduced sunlight and thus photosynthesis, leading to the mass extinctions that seemed to characterize the late Cretaceous period. Thus, from an initial anomaly, the unexpected finding of high levels of iridium at the K–T boundary, a new hypothesis to account for the mass extinctions was born.

Testing the Impact Hypothesis

An obvious way to test the asteroid hypothesis would be to use the sequence of dinosaur fossils in the geological record to trace the time of extinction

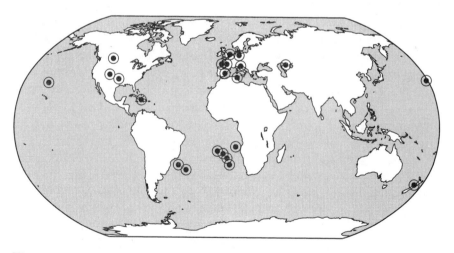

Figure 4.10 Location of the most important sites where iridium anomalies occur at K–T boundaries, tabulated as of mid-1983. The map shows the global distribution of the sites and the variety of environments involved: marine, swamp, and land. The chemical analyses come from laboratories in Switzerland, Holland, the former Soviet Union, and the United States.

with respect to the K–T boundary layer. We can set up a deductive syllogism as follows:

If . . . extinction of the dinosaurs was primarily the result of the meteor impact that also produced the iridium layer,

then . . . dinosaur fossils should be prevalent below the K–T boundary and scarce or nonexistent above it.

One particular fossil bed at which the prediction could be tested was the Hell Creek formation in Montana. Examination of this site, however, revealed that the most recent dinosaur remains were well below the iridium layer—in fact, *too far below!* A full 3 meters (approximately 10 feet) of sediment lay between the last dinosaur remains and the iridium layer. This "3-meter gap," as it was termed, suggested that the dinosaurs had become extinct long before the hypothesized meteorite impact, thus falsifying predictions stemming from the Alvarez hypothesis and rendering it invalid.

In defense of their hypothesis, the Alvarez team argued that the lack of dinosaur fossils in the 3-meter gap did not necessarily mean that the animals had died out at that time but only that they were not fossilized in this specific locality during that period. The logic here is straightforward: The presence of dinosaur fossils *above* the boundary layer would have indicated clearly that the animals continued to live in the area after the impact and thus invalidated the impact hypothesis. On the other hand, given the accidental nature of fossil formation, the lack of fossils in the 3-meter gap did not necessarily mean that the dinosaurs had become extinct. In this case, negative evidence is not as logically binding as positive evidence. (As forensic experts say, "The absence of evidence is not evidence of absence.") The impact hypothesis remained viable, although entirely speculative. Although it was well documented that large meteorites had struck the earth periodically in the past, leaving telltale craters, none of these craters seemed to be large enough nor of the right age to support the Alvarez hypothesis.

A dramatic breakthrough would soon occur. In the early 1990s, geologists discovered the remains of a gigantic impact crater in the Yucatán peninsula and determined its age to be precisely that demanded by the Alvarez hypothesis: 65 million years. Even with this impact event strongly confirmed, however, the paleontological data concerning a mass extinction at that time remained problematic.

Extinctions and the Paleontological Record

One major problem in determining whether a mass extinction occurred involves two intertwined questions. (1) What counts as a mass extinction as compared with the ordinary occurrence of extinctions over a specific time frame? (2) What taxonomic levels should be used in computing extinction rates: species, genus, family, order, or phylum? Both questions involve setting some consistent but ultimately arbitrary criteria.

Determining what counts as a mass extinction may *seem* simple, but it is far from being a straightforward matter. Is a mass extinction equivalent to 90%, 40%, or 20% of *all* species? Should a distinction be made between animals and plants or between terrestrial organisms more vulnerable to drastic climatic changes and marine organisms protected by the shielding effect of ocean waters? Therefore, part of the debate about the meteor impact hypothesis was that different investigators had different criteria for what counted as a mass extinction. For example, at the K–T boundary, 60%–80% of all marine species became extinct but only about 15% of all marine animals. On the other hand, microscopic marine animals (zooplankton) were particularly hard hit, whereas microscopic marine plants (phytoplankton) were not. Organisms living as part of coral reef ecosystems were major victims, but animals that built and lived within the coral reefs survived well. Deep-sea organisms survived better than surface-water organisms, whereas terrestrial plants survived better than terrestrial animals. The selectivity of extinction makes it difficult to decide on which groups of organisms to focus in determining whether an extinction is truly "mass" in character.

At first glance it might seem that species is the obvious taxonomic level at which to examine extinctions. However, using species in the fossil record is not as easy as it might sound. One of the major problems is that it is often far more difficult to draw clear demarcations between fossil species than between living species. With living species, the taxonomist has available evidence not only from anatomy but also ecology, physiology, behavior, and reproductive capacity. Except for certain kinds of invertebrate species that possess easily fossilized parts (shellfish are a good example), most of the characteristics of fossils are obscured so that the paleontologist has much less information to go on. Also, paleontologists often do not have a sufficiently large population of fossils to determine whether their collection represent a single species with many variations or two or more distinct species. Moving up the taxonomic scale to class or phylum includes so many varied forms that tracking each group through time probably would not reveal much fine detail about extinction rates. Most of the species of the class Reptilia, to which dinosaurs belong, might have become extinct, but the entire class might not be considered to be extinct. Thus paleontologists have more or less compromised on the family as the taxonomic level that yields the most useful information about extinction.

Families have enough diversity to withstand various climatic and other changes, yet they are a small enough grouping to reflect more serious environmental events, such as a meteor impact. Family-level differences are also easier to distinguish in the fossil record than are genus- or species-level differences.

However, one problem is inherent in using any taxonomic category higher than the individual species. A family, for example, may contain one genus and species (such as the family Hominidae, which includes humans as the only surviving species), or it may contain hundreds of species, as do some insect families. Yet, single-species families and multispecies families each count as one unit in tallying up extinctions. As a result of this problem, critics of mass extinction theories argue that it is possible to make anything a mass extinction by choosing the taxonomic category that will give the desired results. It is for just this reason that paleontologists negotiated the family as an agreed-upon taxonomic category when trying to assess the severity of most extinction events.

A second problem encountered by the impact hypothesis is that of accurately determining the age of fossils and geological strata. Depending on the dating methods used, estimates of the age of specimens have changed dramatically over the past 30–40 years. To estimate precisely when a species became extinct over the long periods of the geological record it is necessary to use several different databases.

Periodic Mass Extinctions

In 1983, paleontologists David Raup and Jack Sepkoski at the University of Chicago became intrigued with the Alvarez impact hypothesis. In a paper published in the journal *Science* in 1984, they argued that a mass extinction event had occurred not only at the K–T boundary but appeared to be periodic, occurring on the average of every 26 million years.

Such a periodicity seemed remarkable and very surprising. The methodology of determining periodicity, to say nothing of extinction rates, seemed fraught with difficulties. First is the problem of periodicity itself. Probability theorists know that many purely random events often generate the appearance of cycles. Suppose, for example, you take a deck of cards and draw out one card every day. If you do this every day for 250 days, noting on a calendar the days on which you draw a black ace, you would expect to make a mark on the average of twice every 52 draws, or once every 26 days. In reality, however, you do not get such a neat distribution. Five such series of drawings, each spanning 250 days, were simulated on a computer and the distribution of black aces calculated (Fig. 4.11). Note that even though the process of generating the draws is random, black aces appear in

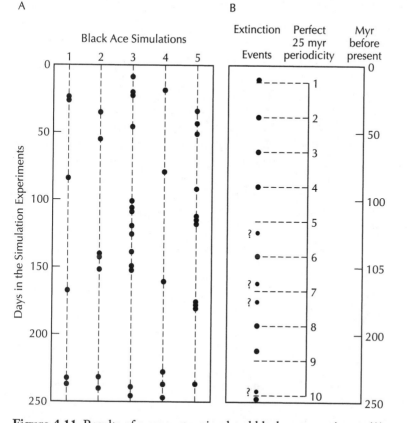

Figure 4.11 Results of a computer-simulated black ace experiment (A) compared with the periodicity established for mass extinction events (B). Myr stands for millions of years. Both series show some degree of clustering, with events bunched together and then separated by intervals in which no events occur. In (A), the periodicity is generated completely by chance and the cycles are somewhat irregular. In (B), the periodicity is more regular, but its cause is not known. The point of the experiments is to demonstrate that even completely random events may show periodicity, suggesting that the mere existence of a cycle does not indicate it is caused by some regular, as opposed to a random, phenomenon.

clusters, rather than being evenly spaced. These clusters are separated by long periods in which no black aces were drawn at all. Here, the periodicity is simply the result of chance. However, Raup demonstrated that the clustering shown in Figure 4.12, the 26-million-year cycle, is more regular than that of the kind of random distribution shown in the example of black aces shown in Figure 4.11, suggesting that extinction events might not be

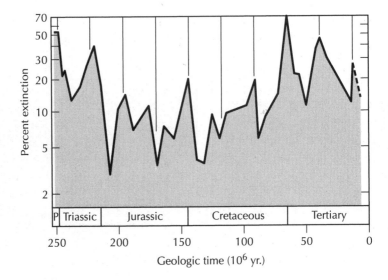

Figure 4.12 Cyclic recurrence of mass extinction events over the past 150 million years for families of extinct marine organisms. The relative heights of the various peaks are approximations. Note that the peaks appear to recur in approximately 26-million-year cycles.

the result of mere chance. In other words, although randomness *might* have generated the 26-million year cycle, its greater regularity argues for a more persistent causal agent. Raup and Sepkoski were thus surprised and delighted to learn that Walter Alvarez and astrophysicist Rich Muller had found a 26-million-year periodicity in the occurrence of asteroid craters and that this periodicity was identical to that found for mass extinction events. A periodicity discernible from three different kinds of data—iridium sampling, dating of meteorite craters, and extinction events—greatly increased the likelihood that these were not purely chance events but had some fundamental connection with each other. Therefore, the Alvarez hypothesis, devised originally to explain only a single extinction at the end of the Cretaceous period, might be extended to a whole series of extinctions occurring regularly every 26 million years.

The Nemesis Hypothesis

After 1984, a number of hypotheses were advanced to explain the periodicity of the extinction and impact events: sunspots cycles, cosmic dust clouds through which the sun passes periodically in its journey around the galaxy, oscillation of the sun above and below the galactic plane, and even the existence of an unknown planet beyond Pluto. For a variety of reasons, none of these hypotheses seemed satisfactory.

One hypothesis, however, remained among the most ingenious, controversial, and enduring: the Nemesis hypothesis. This hypothesis proposed that a dark companion star, Nemesis (named for the Greek goddess who continually followed the rich and powerful, tormenting them), revolved around a common center of gravity with the sun. Nemesis was hypothesized to have a highly eccentric orbit that took it from beyond the reaches of the solar system to the realm of the outer planets every 26 million years. Beyond the orbit of Pluto, Nemesis would pass through the Oort Cloud, a hypothetical ring of millions of comets surrounding the solar system. During this passage, Nemesis would cause immense gravitational disturbances, sending many comets and the debris that make up their heads hurtling through the solar system. It is this material that was hypothesized to impact on earth, creating immense heat, ash, and dust, thereby blocking out the sunlight and, ultimately, causing mass extinctions.

The Nemesis hypothesis is intriguing. It draws on a number of scientific disciplines—geology, paleontology, and astrophysics—and seems to account for data derived independently from them all. The hypothesis has one major difficulty, however. No astrophysicist has any hard evidence that either a dark companion star to the sun or the Oort Cloud actually exist. Before the interest in mass extinction, some circumstantial evidence had suggested the existence of the Oort Cloud, but it has yet to be detected directly. The hypothesized dark companion to the sun and the Oort Cloud are examples of what are called ad hoc hypotheses, that is, hypotheses that explain existing data and possess no independent supporting data of their own but may eventually find support beyond the data they were originally designed to explain. Such hypotheses are useful in science to the extent that they suggest new lines of research. The Nemesis hypothesis has stimulated a number of astrophysicists to begin a systematic search for the hypothesized dark companion star and a more intensive search for evidence of the Oort Cloud.

Such research ventures are often chancy, however. If a nemesis is found, it will constitute one of the great discoveries of this century. If it is not found, we will never know with certainty whether the dark companion star is simply a convenient figment of the scientific imagination or actually exists but is too difficult to detect with our present methods. Thus, questions about mass extinctions and their possible periodic nature are likely to remain unresolved for years to come.

Conclusion

As just suggested, the nemesis hypothesis represents a very different sort of hypothesis from the laboratory and field research examined earlier in this chapter. Nemesis deals with a hypothesized historical event or series of events that cannot be investigated directly, and, therefore, all the evidence is necessarily indirect and circumstantial. The impact crater discovered in the Yucatán is the right size and age for the event that might have caused the extinction at the K–T boundary, but we have no direct way of determining conclusively a cause-and-effect relationship between that event and the disappearance of the dinosaurs and other species 65 million years ago. Furthermore, the Nemesis hypothesis is really not a single hypothesis but a series of separate hypotheses combining observations from the fields of geology, paleontology, geochemistry, and astrophysics. Such synthetic hypotheses certainly play an important role in science. However, although fruitful in stimulating research, it is not possible to expect that such hypotheses will be tested with the same level of rigor applied in laboratory and field studies.

Exercises for Critical Thinking

4.1. Three examples of scientific research have been presented in this chapter:
   a. Laboratory research on nerve growth factor;
   b. Field research on salmon migration; and
   c. Paleontological studies concerning the sudden disappearance of the dinosaurs and other species occurring in the late Cretaceous period.

Remembering our discussion of science historian Thomas Kuhn's concept of paradigm shifts (Chapter 2), rank these three studies in terms of the magnitude of paradigm shift (if any at all) that the results of each might be said to represent.

4.2. In this chapter we noted that Elmer Bueker had found that mouse cancerous tumors grew quite readily in chick embryos. We also noted that Stanley Cohen and Rita Levi-Montalcini found that snake venom was a rich source of nerve growth factor (NGF). Beyond the research on NGF, what might be another significance of this finding?

4.3. Assume it has just been reported in a scientific journal that high levels of iridium have been found in a layer of rock formed in a geologic period in which the fossil record presents no evidence of any extinctions and no evidence of a meteoritic impact on the earth. What would be the effect of such a discovery on the Alvarez K–T boundary hypothesis?

4.4. Using the three case studies in this chapter, discuss how hypotheses were tested by both experimentation and observation. In which case study was observation the most prominent means of hypothesis testing and why?

Further Reading

On neuronal growth

Cowan, W. Maxwell, ed. *Studies in Developmental Neurobiology. Essays in Honor of Viktor Hamburger.* New York, NY: Oxford University Press; 1981.

> *This volume contains some useful interpretive essays, including especially the one by the editor and one by Levi-Montalcini.*

Hasler, Arthur D. *Underwater Guideposts: Homing of Salmon.* Madison, WI: University of Wisconsin Press; 1966.

> *Hasler's original work on the homing instincts of salmon as discussed in this chapter.*

Levi-Montalcini, Rita. *The Saga of the Nerve Growth Factor.* London, UK: World Scientific; 1997.

> *This is a collection of previously published papers by Levi-Montalcini and colleagues on nerve growth factor, from some of the earliest to the most recent. Not easy reading, but stimulating for those who might wish to pursue the saga in greater detail.*

Oppenheimer, Jane M. Ross Harrison's contributions to experimental embryology. In Jane M. Oppenheimer, *Essays in the History of Embryology and Biology.* Cambridge, MA: MIT Press; 1967:92–116.

> *Provides a good background for Harrisons's early work on the self-directed outgrowth of the nerve fiber from the central nervous system.*

On the mass extinction of dinosaurs

Raup, David. *The Nemesis Affair.* New York, NY: W.W. Norton; 1986.

> *This subject has received an enormous amount of attention in recent years. This is one of the best sources in terms of clarity, brevity, and accuracy.*

Raup, David. *Extinction: Bad Genes or Bad Luck?* New York, NY: W.W. Norton; 1991.

> *This is a more general treatment of the issue of extinction. The Nemesis hypothesis is the subject of only one chapter, but this provides an excellent summary.*

Suggested Websites

1. The Nobel Prize in Physiology or Medicine 1986

http://www.nobel.se/medicine/laureates/1986/index.html

> *Includes the original press releases concerning this award given to Stanley Cohen and Rita Levi-Montalcini for their work on the discovery of nerve growth factor.*

2. The Nerve Growth Factor: Thirty-Five Years Later

http://gos.sbc.edu/l/lm/lm.html

Rita Levi-Montalcini's Nobel Lecture, given on December 8, 1986 at Stockholm Concert Hall, Stockholm, Sweden.

3. Growth Factors

http://web.indstate.edu/thcme/mwking/growth-factors.html

Some sense of the revolution that Cohen and Levi-Montalcini's work has spawned can be gleaned from this partial list of biologically important growth factors that have been studied.

4. Salmon Nation: People and Fish at the Edge

http://www.ecotrust.org/publications/salmon_nation.html

Adapted from the book of the same name published by the Ecotrust organization, this site describes the role of salmon in aboriginal cultures, explores the reasons for the salmon's present ecological predicament, and considers the possibility of a sustainable future for people and salmon. The maps showing the distribution of salmon populations are particularly valuable in considering the issue of salmon homing.

5. Salmon and Steelhead Migration: Some Basic Models

http://www.sci.wsu.edu/idea/Salmon/welcome.html

This site provides information and hands-on exercises to demonstrate how scientists study the effects of various environmental perturbations (dams in particular) on salmon migration through the use of mathematical models. Students are given access to data bases and programs that solve differential equations into order to make predictions on how salmon migration is effected.

6. Dinosaurs and The History of Life

http://rainbow.ldeo.columbia.edu/courses/v1001/dinos.2001.html

A detailed syllabus for a course taught by Professor Paul Eric Olsen of Columbia University. The notes for Lectures 13 and 14 (The Lias, Newark, Glen Canyon, and Stormberg Assemblages - Mass Extinction in the Beginning of the Age of Dinosaurs) *are especially interesting.*

The Social Context of Science: The Interaction of Science and Society

The atomic bombs that exploded over Hiroshima and Nagasaki, Japan, on August 6 and 9, 1945, quickly brought World War II to a close. The physicists who had worked on theoretical aspects leading to production of the atomic bomb had some sense of the immense destructive power that atom splitting might unleash. Until it was actually put to military use, however, most had no concept of how they would feel about their roles in developing this new technology. The resulting massive destruction of life and property jolted scientists into a realization they had not encountered before in such a dramatic way. The deployment of the atomic bomb served as a wake-up call for scientists to consider the technological uses to which their theoretical ideas might ultimately be put.

Biologists and medical researchers would soon be confronted with their own versions of this moral dilemma. Revelations of human medical experimentation in Nazi Germany indicated how far some scientists and physicians were willing to go to obtain funds for their research and advance their own scientific or professional careers.

Genetically identical twins, for example, were subjected to deadly injections of drugs or microorganisms to test whether their genetic make-up gave them the same or different susceptibilities to disease. Mendelian genetic theory was used to justify the involuntary sterilization or killing of people designated as being of inferior quality or heredity: Jews, Gypsies, gay men, lesbian women, and persons with mental problems. When such scientific work was exposed at the war crimes trials held in Nüremberg, Germany, after the war, biologists around the world began to wonder "Could it have happened here?"

This chapter explores both sides of the science in society coin: how science affects society and how society often may affect and determine the direction of science. Recognizing the interactive relationship between science and the society in which it is pursued helps to shed light on the complex factors that influence the way in which new scientific ideas develop and may be used or misused.

Science and Technology: The Public Confusion

Watching the evening news on national television, an announcement often is made that the next segment will be turned over to the network's "science correspondent," John Doe. However, John Doe's reports are usually not about science but about a new artificial heart valve or the latest space shuttle launch. Such topics properly fall under the heading of technology, not science.

There is, of course, a connection between science and artificial heart valves and space shuttle launches. These technological innovations are the result of a great deal of theoretical research in physics, chemistry, and biology. So, too, are the many technological marvels we now take for granted, televisions and computers designed by electronic engineers being perhaps the most obvious examples. Modern medicine, too, is a constellation of technologies based upon pure research in the biological sciences. In this case, physicians, rather than engineers, are the technological practitioners. Engineers and physicians are highly skilled professionals who take advantage of the latest advances in science so that they may apply these findings in their own respective professions.

Yet practicing physicians and engineers are not technically scientists in the sense of carrying out research on fundamental concepts. Certainly, engineers and physicians may be scientists when they carry out research on conceptual problems. Indeed, a great deal of scientific research goes on in the laboratories of medical schools and research institutes. In the National Institutes of Health, research is carried out by investigators who may have both MDs (an "applied science" degree) and PhDs (a research degree).

Engineers and physicians also use scientific reasoning in carrying out their professional responsibilities. Much of the diagnostic medicine involved in pinpointing an illness by analyzing symptoms and clinical tests (for example, urinalysis or blood tests) utilizes precisely the inductive and deductive processes described in Chapter 3. So, too, for that matter, do good automobile mechanics. The underlying logic involved in scientific research, far from being limited to scientists, is readily available to all.

Research such as that carried out in medical schools is often called clinical research. This research may involve patients who show certain designated symptoms and who participate in studies designed to learn more about and improve treatment of a specific disease. Clinical research is but one example of applied science, that is, research designed to solve a particular problem. An engineer working on a problem encountered in trying to get a space probe onto the planet Mars is another example of applied research. By contrast, the work of a biologist studying the development of nerve cells in the wings of bat embryos is conducting pure or basic research. This is an intellectual exercise in scientific curiosity with no immediate aim of producing results that may be useful. Ironically, however, the practical benefits derived from such research have proven, historically, to be more useful than those traceable to applied research.

The Social Construction of Science

The scientifically based technology of military weapons designed to kill or the scientific research that produces a new vaccine to cure or prevent an illness demonstrates clearly the diverse effects of science on society. It is far less common to think of society affecting the practice of science. Of course, as noted earlier, when it comes to funds made available by granting agencies for scientific research or the type of research those agencies are willing to fund, societal influences are obvious. At a much deeper level, however, social influences may well impinge on science at every level, influencing the way in which we formulate our most far-reaching paradigms. We have already noted in Chapter 2, for example, that by being socially conditioned to "see" 48 chromosomes, countless observers believed they observed 48 instead of 46 chromosomes. Even as a result of the language we use, our broader social experience inevitably becomes a part of our science. Thus, the cell may be referred to as being similar to a "factory," taking in "raw materials" and synthesizing new "products." The body is likened to a "machine," and aging is referred to as the "wearing out" of body parts. Artificial heart valves and knee joint or hip replacement prosthetics are often referred to as "spare parts." The natural world is labeled "mother nature." Adenosine triphosphate (ATP), a so-called "high-energy compound" involved in powering biochemical reactions, is often referred to as the energy "currency" of the cell. Evolution is described in terms of organisms devising "strategies" for survival, with the evolution of prey–predator relationships likened to an "arms race." Social values, metaphors, and concepts pervade science at the level of theory construction far more thoroughly than many scientists care to acknowledge.

A standard view of science is often what might be termed the "treasure hunt model," in which the scientist is seeking the hidden treasure of nature's secrets. The treasure exists, ready-made and waiting. The scientist's job is merely to "discover" it. This view is associated with an older positivist philosophy of science. In this philosophy, science is completely objective and uninfluenced by social, political, or philosophical "biases" and, therefore, able to describe nature as it really is. In recent years, the positivist view of science has been critiqued by some philosophers and historians of science as being naïve and presenting an unreal picture of the way science works. An alternative way of understanding the scientific process is known as the "social constructionist" view. This view takes as its starting place the idea that we do not so much *discover* nature as we construct *views* of it. This does not imply that ideas are simply made up. All scientific hypotheses must stand up to empirical testing. As just noted, however, in constructing our view of the natural world we may often use ideas, metaphors, analogies, and terminology that reflect both our social and scientific experience. In this way, we construct a view that makes sense to ourselves and others.

The social constructionist view suggests that particular scientific paradigms may reflect the social values of the time and place in which they were developed, because science, far from being neutral, is actually both culturally situated and influenced. To a greater or lesser degree, therefore, both in the topics chosen for investigation and in the construction of paradigms, science may differ from country to country and from century to century. Two examples here will illustrate the way in which social values and assumptions may influence the construction of scientific paradigms.

The Origin of The Origin of Species: *Darwin and the Political Economists*

Charles Darwin was not the first to advocate the evolution of species. He was, however, among the first to propose a mechanism for the way in which this evolution might occur: the hypothesis of natural selection. As we noted in Chapter 1, this hypothesis stated that:

1. More organisms of a species are born than can survive to reproduce;

2. This overpopulation engenders competition between these organisms for available resources, including food, habitats, water, etc.;
3. All organisms vary slightly from other members of their species. Some of these variations are inherited and will be passed on to offspring;
4. Any inherited variation that gives an organism some slight advantage in the competitive struggle for life will tend to be preserved, whereas any variation that confers a disadvantage on an organism will tend to be eliminated. This process of preservation and/or elimination is called natural selection;
5. As a result of natural selection, the physical and physiological characteristics of a species will change over time, that is, they will evolve.

Darwin published his theory in 1859 in a milestone book with a typically long Victorian title, *On the Origin of Species by Means of Natural Selection, or the Preservation of the Favoured Races in the Struggle for Existence.* The second part of the title is revealing. He had been working on the idea behind the book since the mid-1830s. Others who had proposed something similar about the possibility that species might not be fixed and unchanging were, like Darwin, living and working out their ideas in Victorian Britain from the 1830s through the 1880s. It has been suggested, therefore, that the concept of natural selection was a product of the economic, social, and political developments of the period in which industrial capitalism was reaching its peak and the British empire expanding around the globe. "Although Darwin and his family were strong abolitionists, he could not help recognizing that everywhere the British or European colonists expanded they conquered the indigenous peoples. Darwin's expanded title for his book at once explained and, for those who so believed, justified this process. Moreover, this was also the period in which the theory of political economy, detailing the rules by which the capitalist system operated, became well known. The writings of Adam Smith (1750s), Thomas Robert Malthus (1790s) and David Ricardo (1815–1820) laid the foundations for a theory of monetary accumulation. Their works focused on the importance of competition, the consequent struggle engendered by competition, the necessity of innovation to keep ahead of the competition, and the importance of a division of labor. The outcome of the theoretical interaction of all these factors was that some entrepreneurs would succeed and others fail. This was viewed as the natural outcome of free enterprise and considered to be essential to the progress and improvement of society.

A number of historians have pointed to the similarity between Darwin's mechanism for natural selection and the basic economic theories of the time. In 1836, after returning from his 5-year voyage on the H.M.S. *Beagle*, Darwin read ("for amusement," as he said) Thomas Robert Malthus's *Essay on Population*, first published in 1798. Malthus was a country clergyman who was concerned with the increasing poverty visible in England during the first phase of the industrial revolution. In his essay, Malthus made the assumption that populations of people tend to grow exponentially (2, 4, 8, 16, 32, 64...), whereas their food supply tends to grow only arithmetically (2, 4, 6, 8, 10...). Because no reliable census data existed in England, he had no strong statistical support for this assumption, although he did use some limited data from North America. Nevertheless, from his initial assumption Malthus deduced that populations would always tend to outrun their food supply. He concluded that the only way to prevent famine was for those who did not have the means to support large families to practice birth control (abstinence). Malthus was writing only a few years after the French Revolution, and his expressed purpose was to discover the "laws" of human society that would prevent such cataclysmic social upheavals in the future. The cause of poverty, he argued, was that the poor, yielding blindly to the tendency to reproduce, had too many children. Darwin immediately saw the application of this principle to nature, and, as suggested above, it became a cornerstone of the theory of natural selection.

To highlight the general applicability of the social constructionist hypothesis in this case, it is worth pointing to Alfred Russel Wallace, the other naturalist who independently came up with the idea of natural selection. Wallace had read Malthus's *Essay* in the 1850s while recuperating from malaria on a collecting trip in the Malay archipelago (now Malaysia and Indonesia). For him, as for Darwin, the idea of population pressure and competition seemed to provide a perfect mechanism for the way in which evolution might occur. Another contemporary, Karl Marx (1818–1883), although highly impressed with *The Origin* on first reading it in 1860, nonetheless saw immediately that it could be viewed as a projection of Darwin's society onto the animal and plant world:

> It is curious that Darwin recognizes among beasts and plants his English society with its division of labour, competition, opening of new markets, inventions, and the Malthusian "struggle for existence." [Letter from Karl Marx to Friedrick Engels, June 18, 1862. From Donna Torr, ed. *Marx-Engels Selected Correspondence*. New York, NY: International Publishers, 1934 and 1842:128.]

Does the possibility that Darwin may have attained his initial insight into a mechanism of evolution by reading a treatise on political economy necessarily invalidate his conclusion? Certainly not. One advantage of understanding science in its social context is to recognize that constructing nature according to our social models or visions may provide a creative step forward in hypothesis formulation. However, it is also important to recognize that those same social insights can also act as blinders, obscuring our vision of quite different explanations. The notion of competition for scarce resources was valuable in developing the theory of evolution. Yet, until late in the twentieth century, it blinded many biologists to other mechanisms by which evolution may also occur, for example, by cooperative or symbiotic relationships. Viewing science as a social construct makes explicit a process that has been part of scientific creativity from the outset: the use of metaphors, analogies, and models for interpreting nature. By making this part of the process of science explicit, it allows us to examine the metaphors and models we use and determine whether they are helping or hindering our thinking about the phenomenon at hand. The idea of social construction of science neither diminishes the value of that science nor elevates it to a supra-human level. It simply sees science as part of a social fabric to which we all–scientists and nonscientists alike–are heirs.

Homeostasis: Walter Bradford Cannon and the New Deal

In the 1850s, French physiologist Claude Bernard postulated that complex organisms such as vertebrates, especially mammals, have a variety of built-in mechanisms that control their internal environment (*milieu intérieur*), keeping it constant despite changing outside conditions (climate, types of food available, etc.). The mechanism by which this constancy was attained, however, was largely unknown.

During World War I, Harvard University physiologist Walter Bradford Cannon (1851–1945) was assigned to work in Britain and France on the problems of physiological shock that often follow severe trauma. One of the observations Cannon made was that shock involved the breakdown of many of the body's regulatory processes, for example, those that maintain constant acidity or alkalinity (pH), blood sugar levels, body temperature, and others. His subsequent investigations of shock in dogs suggested that most regulatory processes involve the endocrine (hormone-producing) and nervous systems, particularly that part of the nervous system known as the sympathetic system. Both the endocrine and nervous systems operate without any conscious action on the part of the individual. In a remarkable series of experiments, Cannon removed parts of the sympathetic nervous systems of animals and found that, under normal laboratory conditions, the animals continued to function normally. It was only when conditions were changed dramatically (for example, high or low temperatures or injections with high concentrations of glucose) that the lack of regulatory ability became noticeable. Cannon found that the nervous system controls a number of regulatory responses directly. One example is the muscles that open and close blood capillaries, thereby regulating blood flow into peripheral tissues (Fig. 5.1). The nervous system also controls a number of processes indirectly through the endocrine system. An example of this indirect control is that change that accompanies the "fright-and-flight" response of animals to perceived danger.

Cannon called the interactive processes that maintain the constancy of the internal environment "homeostasis." The concept of homeostasis has become a major paradigm of modern physiology. As framed by Cannon, homeostasis was viewed as a dynamic process. Constancy of the internal environment is maintained despite the fact that substances continually enter and leave the body. Homeostasis describes the complex, continual interchange between organism and environment that preserves this constancy. Cannon saw homeostasis as a series of negative feedback processes in which input and output are constantly monitored by the endocrine and sympathetic nervous systems to maintain this physiological balance.

It has been suggested that, as in the case of Charles Darwin, Cannon's may have drawn heavily on societal factors at several levels. The most obvious and direct was that traumatic shock was a process that had been known from antiquity but had been studied primarily because of its wartime occurrence. Had Cannon not been forced to observe the repeated breakdown of physiological control systems among numbers of badly wounded soldiers, he might not have recognized the central role that such control mechanisms play in maintaining normal physiological responses.

At the same time Cannon was developing his paradigm of homeostasis, he also was influenced by a seminar he attended in the late 1920s and 1930s in the Department of Social Relations at Harvard. That seminar was devoted to the work of an Italian economist and social theorist, Vilfredo Pareto (1848–1923), a strong opponent of classical *laissez-faire* economics, who espoused the ability of market forces to control wages, prices, and profits with no government intervention. Pareto was an advocate of a strong, centralized government that would regulate these processes directly and with a firm hand. Pareto's philosophy had drawn considerable attention in the post-World War I period and even more so after the 1929 stock market crash that rocked western capitalist societies. To Cannon, the resulting economic depression of monetary values and the many social processes by which economic behavior was regulated seemed remarkably analogous to the total breakdown of the body's phys-

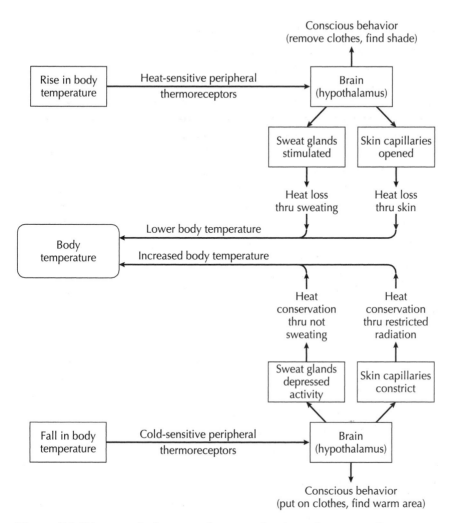

Figure 5.1 Diagram of a homeostatic system for the maintenance of a constant body temperature in mammals such as humans. The human body temperature is kept at approximately 98.6° F (37.5° C), despite changing outside conditions, by a number of components that interact in a series of self-regulating feedback systems. Body temperature, the condition being controlled, is shown in the middle, left. If the temperature rises as a result of exercise or an increase in outside temperature, this increase is detected by receptors in the skin or other parts of the body. A signal then is sent to the brain (to a region known as the hypothalamus), which responds by sending signals to the sweat glands to release perspiration and to the skin capillaries to expand. Both processes produce a cooling effect. Response to falling body temperature works the same way, although in the opposite direction (lower half of diagram). Temperature regulation, as is the case with all other homeostatic systems, works by negative feedback loops. The loops are negative because the body's response has the effect of negating the initial stimulus.

iology during shock. The idea that built-in regulatory processes in government (for example, institutions such as the Federal Reserve Bank, which regulates interest rates and a wide variety of other economic parameters) could stabilize society against future "shocks" seemed eminently reasonable, given the effects of boom-and-bust cycles experienced before and after the 1929 economic depression. Similar thinking motivated Franklin Delano Roosevelt's New Deal programs in the 1930s, in which many regulatory agencies and policies were instituted at the level of the federal government.

Reading Pareto and discussing issues of economic control suggested to Cannon that he might look deeper into the regulatory processes at work within the organism. Although we can never be certain, of course, it is plausible that had he not become interested in Pareto or lived through the economic uncertainties of the postwar period and the Great Depression, Cannon might not have focused his attention on homeostatic processes. Nor might he have formulated an explanation for their means of control in terms of centrally organized regulatory mechanisms. It is certainly possible that this influence may have greatly strengthened his conviction that, in French physiologist Claude Bernard's words, "The maintenance of the constancy of the internal environment is the condition for a free life."

The Social Responsibility of Science

As noted at the beginning of this chapter, many nuclear physicists, including Albert Einstein (1859–1955) were concerned that their work had been used to unleash a weapon of immense destructive power. Similarly, many American biologists were horrified to learn that some of their colleagues in Nazi Germany had conducted experiments on human beings or advocated theories of racial inferiority that were used to justify the enslavement or murder of millions of human beings. These concerns raised issues that had not been widely discussed previously: What responsibility do scientists have for the uses to which their research is put? Can individual scientists be held responsible for the way their work is used? Should scientists inquire into the objectives of those who fund their research, or, as is more traditional, assume that their responsibility ends with making sure their own motives are beyond reproach?

These difficult questions yield no simple answers. The least productive response is to assume that the misuse of science and lack of ethical concerns are products of other times and places and simply "can't happen here." In fact, misuses of science can and do happen everywhere. Two examples will illustrate the moral and ethical issues raised by scientific work in the United States both before and after World War II.

Genetics and Eugenics in the Early Twentieth Century

It was not long after the 1900 rediscovery of Gregor Mendel's work that biologists began finding that a number of genetically controlled traits in human beings appeared to follow basic Mendelian principles. Examples included red–green color blindness, hemophilia, eye color, the basic A-B-O blood groups, and Huntington's disease. In their enthusiasm for these findings, however, some geneticists also claimed that personality and behavioral traits also were determined largely by Mendelian inheritance. Included in this wide range of traits were alcoholism, manic depression, pauperism (the tendency to be poor), rebelliousness, nomadism (the tendency to wander), and "feeblemindedness" (judged by low scores on intelligence tests). Family pedigree studies, in which a trait was recorded in as many generations as the investigator could trace (Fig. 5.2), were used to support the notion that such traits were hereditary. The fact that the trait could be seen to recur in generation after generation was taken at face value to indicate its genetic nature. Because most of these conditions were viewed as negative and because no methods of treatment were known, some geneticists and social reformers argued that individuals who showed these traits (or might be assumed to carry the trait even though they did not show it) should be prevented from reproducing. Such claims were brought together and organized loosely into a reform movement known as eugenics, defined in 1910 by its major American proponent, Charles B. Davenport, as "the science of the improvement of the human race by better breeding."

Using the principles of Mendelian inheritance and family pedigree analysis, eugenicists often made highly simplistic claims that could not be verified by other researchers. Eugenicists thought that they could cure many of the ills of society simply by eliminating defective genes. In fact, they were partly motivated by the prospect of using cutting-edge biology to solve persistent social problems that traditional reform methods had been unable to solve. To achieve their aims, however, eugenicists in the United States lobbied for the passage of compulsory sterilization, to be applied to people judged to have undesirable traits and who were capable of passing those traits on to their children (Fig. 5.3). Between 1905 and 1935, more than 30 states passed eugenics-based sterilization laws that allowed the compulsory sterilization of prison inmates and patients in state hospitals and asylums who were judged to be genetically "defective." In 1933, a major sterilization law, based on the American model, was put into full force in Germany under the National Socialist (Nazi) government, leading to the involuntary sterilization of more than 400,000 people. In addition, eugenicists in the United States lobbied successfully for passage of the 1924 Johnson Immigration Restriction Act. This law selectively

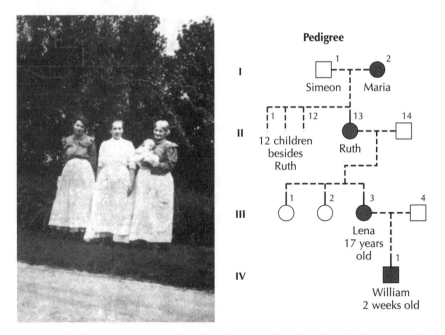

Figure 5.2 A 1937 eugenics "family pedigree" chart for the trait known as "pauperism," a presumed genetic tendency to remain poor. Four generations are shown here, all housed in one almshouse (poorhouse) in upper New York state. In these pedigree charts, squares represent males and circles females. The filled-in squares or circles indicate individuals who have the trait in question, and blank squares or circles are individuals who do not display the trait. As can be seen here, pauperism appeared in every generation shown in the chart, with the latest addition to the family, 2-week-old William (held in the arms of his great-grandmother on the right) already having been diagnosed as a pauper! This chart was drawn up and published in a pamphlet by the Sterilization League of New Jersey.

restricted immigration by establishing quotas on certain ethnic groups (Russians of Jewish descent, for example) and people from countries that were claimed by the eugenicists to harbor genetically defective citizens (Poland, Austria-Hungary, Italy, and others). In the early 1940s, the *St. Louis*, a ship carrying mostly German Jewish immigrants to the United States, was turned back because the Jewish immigrant quota set by this 1924 law had been reached. Forced to return to Europe, many of the passengers would be exterminated in Nazi death camps. In all countries that passed sterilization or immigration laws at this time, it was usually the poorest and most defenseless individuals who were the targets.

Eugenics as both theory and practice did not escape scientific criticism. Nobel laureate Thomas Hunt Morgan (1866–1945) was the geneticist at Columbia University who had introduced the fruit fly (*Drosophila*) as

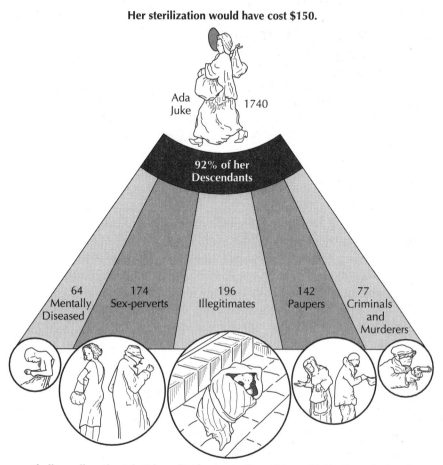

Figure 5.3 A 1937 cartoon calling for the sterilization of "degenerates" to prevent them from passing on their "defective genes" to future generations. The argument brought to the public was that too much tax money was being spent on supporting defectives who should never have been born. Ada Juke was an eighteenth-century woman who had a number of offspring, many of whose descendants became wards of the state. The economic message here was clear: If Ada Juke had been sterilized in 1740, none of the hundreds of defective offspring she produced would have been born and the state would have been spared millions of dollars in needless expense. This cartoon appeared in a pro-sterilization booklet published by the Sterilization League of New Jersey.

the model organism for genetic studies. He pointed out in 1925 that many of the behaviors eugenicists claimed were determined genetically were so vague and ill defined that it would be impossible to demonstrate any significant genetic components. Morgan also noted that the family pedigree data on which eugenicists based their analyses were obtained in such haphazard and uncontrolled ways that they were completely unreliable. In 1932, Morgan's student, Hermann J. Muller, then at the University of Texas and a future Nobel laureate himself, emphasized that until socioeconomic factors were equalized for all people there was no way to distinguish the effects of biology from those of the environment in assigning causes to social behavior. Given the economic disparities that existed after the 1929 depression, Muller argued, eugenics would only be used by the rich and powerful against the poor and weak. By the time such criticisms were gaining force, however, eugenics laws of various sorts had already been passed and the damage had been done. Many of the sterilization laws were not repealed until the 1960s.

The issue of social responsibility here lies with those eugenicists who, in their enthusiasm to apply the latest findings of science to the solution of social problems, ignored the human element in their studies. Eugenicists assumed that the people they designated as genetically defective were less than fully human and should never have been born in the first place. The Nazis, of course, carried this attitude to the extreme, first sterilizing those they claimed were defective and then annihilating those labeled as "lives not worth living."

Herbicides in Southeast Asia

The presence of U.S. military troops in South Vietnam in the early 1960s was officially aimed at preserving the separation of that country from North Vietnam, a division created after World War II. During the course of the Vietnam war, the U.S. Air Force sprayed "defoliants" over 5 million acres of South Vietnamese forests and croplands. As a result of this spraying program, 35% of South Vietnam's dense tropical forest was defoliated, leading to a significant reduction in plant and, consequently, animal life. Spraying of the coastal mangrove swamps resulted in total annihilation of the vegetative cover. The chemical agents used, principally 2,4-D (2,4 dicholorophenoxyacetic acid) and 2,4,5-T (2,4,5-trichlorophenoxyacetic acid), in mixtures known as "Agent Orange" and "Agent White," are called "defoliants" because they cause plants to lose their leaves. Depending on strength and frequency of spraying, defoliants may kill the plants. The military purpose of such massive spraying was two-fold: (1) to reduce the dense vegetative cover of the forested regions so that enemy troop movements could be more easily detected, and (2) to destroy crops in agricultural areas designated as enemy strongholds, thereby destroying the ability of the affected regions to feed the population, a strategy referred to by critics as "ecocide."

How does this relate to the issue of scientific responsibility? Arthur W. Galston, a plant physiologist at Yale University, pioneered research into the nature of plant growth hormones from the 1930s through the 1950s. Among the processes he studied were hormonal interactions that led to formation of the abscission layer at the base of plant leaves where the leaf stalk joins the stem. Triggered by hormones similar in chemical structure to 2,4-D, formation of this layer normally causes leaves to fall off of trees in autumn. In the mid-1960s, Galston was horrified to find that his chemical studies on the hormonal control of abscission had formed the basis for developing Agents Orange and White. He had originally hoped that his work might be used to produce more effective weedkillers for agricultural use. After learning about the destructive purposes to which his research had been put, Galston became an active opponent not only of the military's defoliation policy but also of the Vietnam war itself. He realized, only after the fact, that *scientists generally have no say concerning the uses to which their work is applied.* In the aftermath of the defoliation program, Galston lectured and wrote widely about the need for scientists to take more responsibility for they way in which their work will be used.

The Galston case emphasizes that, although it was impossible to predict what the uses of any new research might be, it was important for scientists to give at least some thought to this issue rather than assume naively that science always leads to beneficial applications. Like the geneticists who decried eugenics, Galston voiced his criticisms only after the damage had been done. His case came to demonstrate, however, that it is the political and social contexts in which research is carried out and used, rather than the research itself, that often determine whether the results are desirable or undesirable.

Ethical Concerns in Science Today

Modern examples of ethical questions that arise in scientific work abound. As geneticists were working fast and furiously to sequence the human genome, questions already were being raised about who should have access to that information. Can insurance companies refuse coverage of (or employers jobs to) people with genetic diseases? Should geneticists encourage the testing of newborn infants for genetic diseases for which no therapies, let alone cures, are available? Such questions are not limited to genetics. Should a researcher working on bat sonar navigation accept research funds from the military, not knowing how that work eventually will be used? If a young PhD is having difficulty finding a job, should he or she accept a job with a biological weapons unit?

One of the most controversial issues to emerge from new technologies surrounds the prospect of cloning animals. Cloning is a process in which stem cells are implanted in the female reproductive tract and allowed to develop. "Dolly," a sheep that was successfully cloned in 1998, represented one of the first of these new technological breakthroughs. Questions arose immediately over the moral and ethical implications of cloning. A company was formed in New York to use stem cells from pets to clone replacements when beloved pets died (the company was named "Copy Cat"). What are the implications for the cloning of human beings? Suppose, for example, a child is killed. Would cloning that child be legal and ethical as a replacement for the grieving parents? What about cloning human bodies for organ replacement? These are very serious issues and raise a host of problems. None are easy to resolve, but difficult questions must be asked now. Means must be devised to guide us toward making ethical and humane decisions in the future. Such issues cannot be put aside if scientists are to take any responsibility for preventing future misuse of their research.

Ethical and Social Issues in the Use of Human Subjects for Research

One of the most startling revelations of the 1945 Nüremberg trials was the extent to which German scientists, supported by grants from the government, had carried out brutal forms of human experimentation. Subjects were placed in ice-cold tanks until they died, an experiment designed to assess survival of pilots downed in the icy waters of the North Sea. Other subjects were given poisonous gases in an effort to determine strategies for military personnel in gas warfare. Others were used for transplantation of various organs, at a time when such procedures were merely speculative experiments. The list is distressingly long.

The issue of human experimentation has always raised difficult ethical questions. Clearly, if we are to make any medical progress, at some point new medications or new surgical procedures have to be tried on *somebody*. The questions remain: How much risk is ethically acceptable, and how much benefit will come to the subjects of such experiments? No simple formula answers these questions, but it is critical that scientific investigators keep ethical considerations forefront in their minds as they plan clinical trials that involve human subjects.

The Tuskegee Study

The fact that scientists and medical personnel do not always carefully consider the ethical implications of their work is illustrated by a project conducted between 1932 and 1972 by the United States Public Health Service (USPHS) in Macon County, Alabama, a rural, cotton farming area. The project was called the "Tuskegee Study," after one of the larger towns in the area, home to the renowned Tuskegee Institute founded by George Washington Carver to promote higher education for southern blacks. The purpose of the study was to trace the long-term effects of syphilis infection. Syphilis is a sexually transmitted disease caused by a spirochete bacterium. If infected persons are not treated, they may harbor the parasite for the rest of their lives, which are often shortened considerably as a result. The spirochetes may invade virtually all kinds of body tissues, producing skin lesions, debilitating pain in the joints and muscles, liver deterioration, and invasion of the central nervous system and brain, resulting in paralysis and insanity. Spirochetes may sometimes remain dormant in the human body for years before reemerging and continuing to spread throughout the body. In other cases, the organism has no latent period, moving from one part of the body to another immediately after infection. Clinically, syphilis is usually described as passing through three stages: primary, secondary and tertiary, the latter representing the final and usually fatal stage.

Although syphilis had been a long-standing problem in the United States, it became a major public health issue during and after World War I. Attempts to study the problem systematically were hampered by strong religious and moralistic attitudes that shunned any open discussion of sex or sexually transmitted diseases. As a result of lobbying by religious groups, congressional appropriations to the USPHS for such studies were continually cut during the 1920s and were even more drastically reduced after the stock market crash of October 1929. One of the questions doctors needed to know in order to learn how to treat syphilis was how it progressed over long periods of time if left untreated. Why was it, for example, that some people progressed immediately from one stage of infection to another, whereas others showed varying periods of latency? Were there differences in the strains of spirochete or differences in the people? Could knowledge of those differences provide a clue to possible treatment?

In order to answer these questions, the USPHS initiated a long-term study in 1932. It was generally assumed (although with no hard statistical evidence) that syphilis was especially rampant among African-Americans. Indeed, in some medical circles African-Americans were referred to as a "syphilis-soaked race," an indication of attitudes prevalent both inside and outside the medical community. A project was therefore conceived to follow a syphilitic population of 399 rural African-American men during the long-term progression of syphilis. The project was managed out of the

Centers for Disease Control and Prevention (CDC) in Atlanta, and involved a number of local doctors as well as interns and medical officers from the USPHS in Washington. African-American nurse Eunice Rivers was hired to act as a local contact with the subjects and as an intermediary between the white doctors and the black community.

USPHS officials, with the help of Nurse Rivers, went to the fields and homes of rural sharecroppers in Macon County, talking to individuals and selecting for the study those who could be diagnosed in the tertiary stage of syphilis. Potential subjects were told they would get periodic free medical exams, a hot lunch on the day of each exam, transportation to and from the examination site, and burial expenses (a rather macabre but economically attractive incentive). They were also told they would receive treatment for various illnesses. No one, often not even the young interns assigned to the project, always knew what the real purpose of the study was. Many of the subjects thought they were being examined for rheumatism or "bad stomachs." Nurse Rivers handled most of the negotiations and the day-to-day work of chauffeuring subjects to and from the study.

The control group included 201 men who were free of the disease and who were examined regularly by USPHS doctors. Because the point of the study was to examine the long-term effects of syphilis, there was never any intention of offering treatment. Some subjects were given placebos, fake or "empty" pills that subjects were told was a "cure." The role of placebos, legitimate in much medical research, is to ensure that improvement in a condition is really because of the treatment administered and not a psychological effect of the subject *believing* he or she is being cured. Reports of the Tuskegee study were made in major medical journals periodically during the course of the study. Thus, anyone who read the literature carefully should have known exactly what was going on in the Alabama study. No attempts were made to conceal the study or keep the findings secret. Approximately 100 (25%) of the 399 men with syphilis in the study died directly from the effects of the disease.

The whistle finally was blown on the Tuskegee study in 1972, when issues of civil rights were at a new high. The first exposure of the study came from a reporter at the Associated Press. Other newspapers quickly picked up the story. Under pressure from the press, USPHS officials (of the initial planners, only Nurse Rivers remained with the project) sought in vain to find a written protocol for how the study was designed or for any form showing that subjects had given official or informed consent for their participation. Responses were immediate and angry, especially from the African-American community. Although most newspapers stopped short of calling the project a Nazi-like experiment, some African-Americans considered it another example of genocide. The Tuskegee study was abruptly terminated, but the questions it raised stimulated a major reassessment of USPHS protocols for studies involving human subjects. The National Institutes of Health, which provides funding for investigative work, now require elaborate reviews of all projects in which human subjects are involved, not only to protect the subjects' health and well being but also to insure that privacy and other matters of personal concern are respected.

The Tuskegee Study raises serious questions about the ethics of human experimentation. In the first place, the doctors conducting the study deliberately withheld treatment from individuals who, had they been given available medication, might have benefited or at least might have experienced less suffering. By the 1950s, penicillin was available widely and known to help reduce syphilitic infection. Yet the whole project was conceived with the purpose of making observations on untreated individuals, an issue of enormous ethical implication that was apparently not seriously considered by any of those who designed or participated in the study. That the study was carried out by medical personnel makes the question even more problematic. A part of the Hippocratic oath, which all physicians take upon being admitted into the medical profession, states that the physician must be dedicated to "doing no harm." Those carrying out the study were not the only ones at fault however. More than a dozen peer-reviewed journal articles on the Tuskegee study appeared in the medical literature over a 40-year period, with no questions raised about the ethics of the study design, especially the issue of withholding treatment. The moral question simply did not seem to exist, as it had not existed for many physicians in Germany under the Nazi regime.

The Tuskegee study also revealed the attitudes of the USPHS and its staff. The very fact that it was assumed that most black men had syphilis and that, therefore, they would constitute a ready-made population for such investigations indicates the deep racist biases that underlay the study design. Black physicians knew that syphilis was not rampant in their communities. In the 1930s and 1940s, however, African-American doctors were not admitted into the American Medical Association (AMA) and had little chance to interact with or explain to their white counterparts the lack of basis for this myth. Furthermore, the use of poor blacks from the rural south presumed that no one would question the study or raise the same kind of questions that would be raised if the study were to be conducted on middle-class white men (this was insured further by keeping the real purpose of the study secret from the subjects). That the experimental group was an impoverished and desperate segment of American society made them all the more likely subjects. As one reporter put it, the fact that these men agreed to participate because they got a free exam, a hot lunch, and burial expenses shows just how marginal their life was at the time.

Medical researchers have learned much from the exposure of the Tuskegee study and other cases where human experimentation has been carried out with less-than-sound ethical considerations. Yet neither scientists nor the public can become complacent and assume that something like Tuskegee will never happen again. The right to investigate must always be balanced against the rights of the individuals participating in any given study. Scientists can easily get so caught up in enthusiasm for their research that they simply do not think about the ethical questions involved or about their subjects as human beings with fears, rights, and concerns. Ethics are not a philosophical hobby that scientists can think about in their spare time; they must become continuous and rigorously monitored aspect of the research process.

Science and Pseudoscience

Pseudosciences (Greek, *pseudes*, false) are activities or forms of thought that strive to be identified as scientific yet lack strict adherence to the intellectual structure that characterizes science. A listing of all the pseudosciences that have appeared over the course of human history would fill volumes. Phrenology, for example, was a nineteenth-century pseudoscience that claimed a correlation between the shape and contours of a person's cranium and his or her personality traits or illnesses. The ancient art of palm reading is yet another pseudoscience and purports to read the past and predict the future by tracing certain lines found in the palm and fingers of the hand.

Perhaps best known among modern pseudosciences is astrology. Daily newspaper columns devoted to astrology are widely read. In this pseudoscience, a supposed correlation between the date of a person's birth and the changing alignments of the planets or other astronomical bodies leads to formulation of advice on topics as disparate as whether to invest money, become romantically involved, or find a new job. Astrology has a wide appeal among many segments of society—even residents of the White House have been known to seriously seek the advice of astrologers! However, astrology, like phrenology and palm reading, has failed consistently to show any statistically significant degree of predictive accuracy other than would be expected by chance alone. This relegates it to the realm of a relatively harmless pseudoscience. In the next section, however, we will deal with "Scientific" creationism, a far from harmless pseudoscience that has been able to exert considerable political power.

"Scientific" Creationism

In August, 1999, the State of Kansas Board of Education voted to remove the requirement that evolution be taught in biology classes in the public school curriculum. References to the concept of evolution of new species from older forms (macroevolution) were to be eliminated from classrooms and texts. The Board also voted to remove from the curriculum the currently accepted "big bang" theory about the origin of the universe. This vote occurred despite letters from the presidents of the state's six public universities, warning the Board that the new standards would "set Kansas back a century and give hard-to-find science teachers no choice but to pursue other career fields of assignments outside of Kansas." Only a June 2000 vote, ejecting two of the three "Creationists" on the Kansas State Board of Education, appears to have reversed matters in terms of educational policy in that state.

Yet what happened in Kansas has happened in many other places in the United States. Attempts to either have Creationism taught as a scientifically acceptable alternative hypothesis to evolution in high school biology classrooms or to eliminate evolutionary theory altogether have taken legal form in Tennessee (1973), Minnesota (1978), Louisiana (1981), and Arkansas (1981). Although most of these measures failed to pass in state legislatures or were immediately challenged and found to be unconstitutional, the Creationists' efforts clearly represent the deep-seated anxiety that, historically, paradigm shifts have often elicited. As Thomas Kuhn emphasized, even within science, such shifts do not come about easily.

Some Historical Background

The vote of the State of Kansas Board of Education is a modern example of the intellectual conflict between science and religion that has been going on for more than 400 years. It had its beginnings with the publication of Nicholas Copernicus's (1453–1543) book, *De revolutionibus orbium celestium* (Latin, On the revolution of the heavenly bodies). In this work, Copernicus rejected the old astronomical paradigm of an Earth-centered universe, claiming instead that the sun was at the center, with Earth and the other planets orbiting it. The struggle for acceptance of the Copernican system, especially the confrontation between Galileo Galilei (1564–1642) and the Roman Catholic and Protestant churches in the early seventeenth century over advocacy of the new system, is well known and has been described in a number of sources (see the reading list at the end of this chapter).

A similar version of that confrontation occurred over evolution in the nineteenth and twentieth centuries. At the annual meeting of the British

Association for the Advancement of Science in Oxford, England, in 1860 (the year after *On the Origin of Species* was published), the Reverend Samuel Wilberforce, Bishop of Oxford, ridiculed Darwin's theory. He asked Thomas Henry Huxley (1825–1895), a young naturalist and Darwin supporter, whether he was descended from an ape on his grandmother's or grandfather's side! Huxley, who had only been persuaded to attend the meeting at the last minute, replied that he would rather be descended from an ape than from a "man of high position" who used his talents to such poor effect. From this point onward, the "warfare between science and religion" (to use Huxley's phrase) focused on the theory of evolution, especially human evolution. The debate continues today with great intensity, although almost entirely in the United States. The current form of the controversy centers on the issue of evolution versus creationism, and the arena has been the public school curriculum.

Creationism: What Is It?

Creationism is the belief of fundamentalist Protestant Christians that the creation of the earth and all life upon it occurred in 6 days, precisely as described in the book of Genesis. Estimates of how long ago this event occurred tend to vary from one group of creationists to another. Some support the calculation based upon the ages of the prophets in the Hebrew Bible (Old Testament) by Archbishop James Ussher (1581–1656) stating that these fateful 6 days occurred 4004 years BCE. Others seem willing to accept a considerably older earth without life on it, but hold steadfast to the belief that the 6 days of creation of life could have occurred no longer ago than 6000 years. Still others now seem willing to accept a figure as high as 25,000 years. The precise number does not really matter, because all fall far short of geologists' estimates of the earth's age. As for living organisms, creationists maintain that God created each species as separate and distinct groups. Although most creationists are now willing to grant that some variations may be found *within* species (microevolution), they deny that new species may emerge from older species over time (macroevolution). Thus, it is the concept of the evolution of new species that is the main focus of their attack.

The vote of the Kansas State Board of Education described at the start of this section marked a return to a policy, that seemed to be one step closer to the once-widespread practice in the United States, of prohibiting the teaching of evolution. In a famous 1925 trial in Dayton, Tennessee, science teacher John Scopes was accused of violating a state law designed to enforce this prohibition. In this face-to-face contest between science and religion,

Clarence Darrow (1857–1938) argued for the defense and fundamentalist William Jennings Bryan (1860–1925) argued for the prosecution. The trial received massive publicity, becoming the first to be broadcast nationally by radio. The result, as reported by the press, was that Darrow, who knew as much or more about the Bible as Bryan, "made a monkey of the man." Scopes was convicted—his "guilt" in teaching evolution was never denied—but fined only a token $100, which was never paid. The courtroom confrontation, which was later the subject of a Broadway play (*Inherit the Wind*) and at least four films, would forever be known as "the monkey trial."

Educational change was slow in coming. In fact, the net effect of the trial was that evolution, which had previously been included in high school biology texts, was dropped by all major publishers after 1925. This omission continued through the early 1960s, when massive increases in the amount of federal aid to science education were aimed at raising the level of science education in the United States, and evolution gradually assumed a central role in modern biology education. The Tennessee law under which John Scopes had been convicted finally was repealed in 1965.

Before the Scopes trial and even after, Creationism most often was seen from inside and outside for what it was: a religious view held by fundamentalist Christians. However, after World War II, the reintroduction of evolution into school curricula (if not yet textbooks) led creationists to change their tactics. On April 30, 1953, Senate Bill 394 became law in the State of Tennessee. Unlike the earlier Tennessee law, this did not prohibit the teaching of evolution but required that "scientific creationism" be presented in the biology classroom as an alternative scientific hypothesis to the Darwinian view. Similar legislation was considered in several other states. In 1980, a law resembling that of Tennessee and mandating the teaching of creation "science" was passed as Act 590 in the Arkansas legislature.

"Scientific" Creationism Versus Evolution

On the surface, the idea of teaching "scientific" creationism side-by-side with the teaching of evolution seems like an acceptable compromise, if only in the spirit of fair play for opposing viewpoints. Unfortunately, doing so presents several problems. The first of these problems supersedes the others. Creationists present no testable paradigm of their own. Instead, their tactic has been to criticize aspects of the supporting evidence for modern evolutionary theory.

There is absolutely nothing wrong with criticizing an existing paradigm, of course. Scientists criticize each other's ideas all the time; this is an essential part of the scientific endeavor. In attacking evolutionary theory, one of the primary targets of the creationists has been the existence of gaps in the fossil record between earlier forms of species and their descendants. Because it takes a rare combination of environmental factors for fossils to form (to say nothing then locating them!), one would *expect* the fossil record to be incomplete. Furthermore, some organisms are better candidates than others for fossilization (for example, animals with skeletons in contrast to those without, such as jellyfish). Despite these problems, however, the existence of transitional forms between groups of animals and plants is far more complete than creationists like to admit (Fig. 5.4). As Eugenie Scott, Executive Director of the National Center for Science Education has put it, "there are [many] transitional fossils....the problem is [antievolutionists] will never tell you what they would accept as a transitional form." No matter how many are discovered, the total will never be enough to satisfy the creationists, who can always point to whatever gaps may remain.

Yet another point often raised against evolution by creationists is that living organisms, including humans, are far too complex to have evolved by chance. Consider, for example, the following typical examples from a creationist book:

> ... will the [chemical] elements of the earth, left to themselves, ever produce an automobile, or even a simple gear? To the contrary, the elements remain as they are. [From a contemporary volume published by the Watchtower Bible and Tract Society. Brooklyn, NY: International Students Association; no date.]

Quoting from the same creationist source:

> Take a large barrel and put into it bits of steel, glass, rubber and other materials. Turn this barrel thousands of times and open it. Would you ever find that the materials by themselves had produced a complete automobile? [From a contemporary volume published by the Watchtower Bible and Tract Society. Brooklyn, NY: International Students Association; no date.]

The answer to both questions is obvious: Of course the elements or pieces of a car would not come together automatically, nor would any biologist claim otherwise. However, according to the Darwinian paradigm, simple molecular components or "pieces" of the organism do not come ready-made but evolve together as part of the whole organism. Little by little, these parts have been shaped in conjunction with one another by natural selection. Indeed, it is the action of natural selection that keeps evolution from being a strictly chance or random affair (although the mechanism of most genetic change, mutations, *is* a random matter). By arguing as they do in the quoted passage, creationists use the tactic of setting up a straw

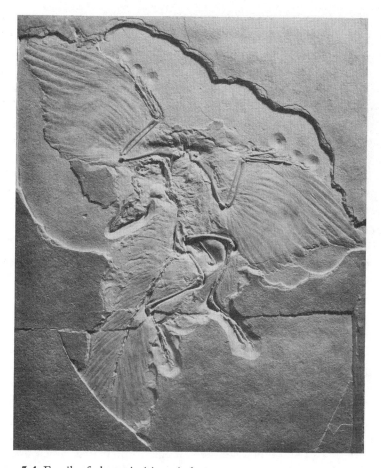

Figure 5.4 Fossil of the primitive *Archaeopteryx*, a true intermediate between modern birds and reptiles. Note the long tail, claws on the appendages, and teeth, all of which are reptilian characteristics, yet the perfectly formed feathers and front limb structure are characteristic of modern birds. Although *Archaeopteryx* is not considered a direct ancestor of modern birds (it shared a common ancestor with a form that did give rise to modern birds), it clearly represents the sort of intermediate form from the fossil record that Darwinian theory would predict. In fact, without the feathers, *Archaeopteryx* would probably be classified as a small, bipedal dinosaur. [© F. Gohier/Photo Researchers.]

man argument, one that serves only to demonstrate their own complete lack of understanding of the evolutionary process.

As noted earlier, creationists admit that change may occur *within* species, but maintain that selection has never produced divergence from one species into two *new* species. They note that the many varieties of dogs, the result of selection by humans, are all still the same species and can interbreed. Because, in nature, the evolution of most species has required long periods of time, far longer than any age of the earth accepted by most creationists, this is obviously a difficult objection for biologists to counter. Yet, genetically isolated populations of fruit flies have been produced in the laboratory by selecting two different lines and keeping them apart for many generations (14 months), demonstrating that the basic barrier between species, reproductive isolation, can form in a relatively short period of time.

Another area of attack by creationists is on the methods used by geologists and paleontologists to estimate the age of the earth and that of fossils and the geologic strata in which they are found on the earth's surface. Creationists point to the fact that estimates of the earth's age have changed radically in the last century, from somewhere less than 3 billion to the current estimate of about 4.5 billion years. With such constant changes in the figures quoted, creationists argue that the methods of estimating the age of the earth or fossils are so faulty as to be unusable. Here, they *totally* miss the nature of scientific investigation: to continue to learn new ideas and obtain more precise data. The revision of ideas marks a strength of modern science, not a weakness.

Creationists also claim that because different radioactive isotopes (carbon-14, uranium-238, etc.) give different readings from the same rock sources or fossils, the methods are unreliable and cannot be used. As advocates of a "young Earth" theory, creationists have a stake in discounting dating methods that emphasize the vast age of the planet. Their argument is based on a misunderstanding of the nature of radioactive dating methods. It is true that results sometimes vary, because fossils or rock samples may contain secondary isotope deposits (washed onto the sample perhaps after it was exposed to the air in its natural setting) and some samples contain only small amounts of isotopes, introducing the possibility of measurement error. However, geologists can check one radioactive dating method against another and against an independent method, such as geomagnetic data, comparing the orientation of metal particles in a sample to known orientations of the earth's magnetic field at different periods of geological history. Such comparisons show a high degree of agreement between age estimates using the different methods.

Creationists also have claimed that the weakness of evolutionary theory is apparent in the disagreements that often surface among biologists about specific issues, such as the rate of evolution. For much of the first half of the twentieth century, most evolutionary biologists agreed that the evolutionary process was slow and constant, leading to gradual change over time. In the 1970s, however, some evolutionary biologists suggested from both genetic and paleontological data that species may evolve by alternating periods of rapid evolutionary change followed by longer periods of little or no significant change, called periods of stasis. This theory, often referred to as "punctuated equilibrium," was hotly debated between 1975 and the mid-1990s. Creationists jumped on the bandwagon and claimed immediately that such debates showed the incoherence and weakness of the Darwinian paradigm. In fact, it shows just the opposite. One of the greatest strengths of science is that it fosters questioning of even the most established ideas and brings them up for scrutiny. The debates on punctuated equilibrium helped clarify many questions about evolutionary mechanisms and rates of change that otherwise might never have come to light.

Creationism as a Science

Because creationists have attempted to claim that creation theory is a science, it is worthwhile to examine how it measures up to other sciences. Keep in mind that to qualify as a scientific hypothesis there must be some way to test predictions deriving from the hypothesis.

One problem in examining the creation hypothesis is that of deciding *which* of the creation myths described in the book of Genesis is the proper creationist hypothesis to test. In Chapter 1, after creating all the other forms of life, God creates both Adam and Eve simultaneously: "Male and female created he them." Chapter 2, however, describes a quite different creation myth. In this version, the creation by God of "every beast of the field and fowl of the air" occurs *after* Adam. Later, God is described as putting Adam into a "deep sleep," during which Eve is formed out of one of his ribs. [This second creation myth gave rise to the erroneous belief, once held by many fundamentalist Christians, that human males have one less rib than females.] Religious and ancient history scholars have noted that both stories have different origins and are variations of far older creation myths to be found in ancient Babylonian and Egyptian religions. Because the Hebrew Bible relates in Exodus how the Israelites lived for many generations in ancient Egypt, it is hardly surprising that they would incorporate variations of Egyptian creation myths into their own accounts.

However, let us take just the first story of creation appearing in the book of Genesis and test it, as the creationists insist that we do, as if it were a scientific hypothesis. Doing so is not completely a lost cause for creationism. In fact, this "Genesis I" hypothesis may be said to generate predictions just as accurate as those of evolution by natural selection. For example, the order in which God creates the various species in the hypothetical 6 days of creation is a rough approximation of the order in which the fossil record shows them appearing in the evolutionary scheme of things (although the whale, a mammal, is listed with fish, a common and quite understandable error during the time the book of Genesis was written). The anatomical similarities between humans and our closest relatives, the great apes, as well as those between other groups of species, may be said to be predicted by both the evolutionary and creationist hypotheses, the latter accounting for these similarities on the quite reasonable grounds that God simply used similar plans in making the various kinds of organisms.

However, if creationism is a science worthy of being taught in a science classroom, it must be willing to play by the rules of science. Thus, for example, the theory of evolution by natural selection requires a very long period of time for the evolution of the millions of species we see around us to occur, as well as for those many million more species now extinct. The creationist hypothesis, on the other hand, holds to the 6-day creation period described in Genesis, postulated to have occurred no more than 6000 years ago on a relatively young earth (as noted earlier, ranging from approximately 4000 to 25,000 years in age; recently, some creationists have admitted to the possibility of longer time periods and that the 6 days of creation may not have been literally 24-hour days, a position long held by liberal Christian theologians). Here, however, the evidence from geology, chemistry, and physics is quite clear: The ages of the oldest rocks yet to be found on earth, as determined by a number of different geological, chemical, and physical means, show the age of the earth to be approximately 4.6 billion years, with evidence for the earliest forms of life appearing in ancient geological formations approximately a billion years later. It should be stressed that the geological, chemical, and physical principles and techniques supporting modern evolutionary theory *are themselves based upon the acceptance of theories central to those other scientific disciplines.* Therefore, when creationists deny the validity of contemporary evolutionary theory, they also are denying the validity of these other scientific fields as well. For example, if modern biology is to be taught without its central unifying theme of evolution, so also must chemistry be taught without the atomic theory, physics without quantum mechanics, geology without the principles of age-dated sedimentation and rock stratification, and so forth. All would have to give way to the creation myth described in Chapter 1 of the book of Genesis!

As we stressed in Chapter 2 of this book, science tests its hypotheses by attempting to disprove them, either by designing experiments or making further observations. This being the case, would it be possible for evidence to be presented that would disprove evolutionary theory, thereby forcing biologists to look elsewhere for an explanation for the origin of species? Absolutely. Would it be possible to produce evidence that disproves the creationist hypothesis? Or, putting this question into the scientific mode of expression we have used in this book, with its inconsistent age of the earth and other false predictions stemming from the creationist hypothesis, does that mean it is therefore abandoned as being false by creationists? It does not. In fact, the creationist tactic for dealing with the incompatibility of their view of the age of the earth, based on the Genesis I hypothesis, with that derived from geology and radiochemistry, is highly revealing. As we noted earlier, the creationists on the Kansas Board of Education demanded removal of any mention of the "big bang" theory of the origin of the universe (estimated by astronomers to have occurred between 10 and 20 billion years ago) as well as any current scientific determinations about the age of the earth. By so doing, they would prevent Kansas students from learning about the false predictions deriving from the creationist hypothesis by simply removing those predictions from the curriculum!

Writing a review in criticism of a quite different sort of pseudoscientific hypothesis (one dealing with a supposed genetically based predisposition to commit rape!), Dr. Jerry A. Coyne, of the Department of Ecology and Evolution at the University of Chicago, noted that such theses lack:

> ...the defining property of any scientific theory—the property of falsifiability, the ability to be disproven by some conceivable observation. An unfalsifiable theory is not a scientific theory. It is a tautology, *an article of faith*. [italics ours; *New Republic* 2000:(April 3):29.]

In stating to the Kansas State Board of Education that "The Bible is the sole authority on creation" the creationists nicely confirmed Coyne's words. Creationism is a religious teaching, not a scientific hypothesis. It therefore remains immune to scientific analysis. Most certainly creationism might be taught in schools along with other religious views, but it should be presented as a religious teaching and *not* as science in a biology class.

It is interesting to note that, in their comments to the Kansas State Board of Education, the creationists themselves made the case for the unscientific nature of creationism by stating that evolution and morality were incompatible. Science, of course, does not and, indeed, by its very nature *cannot* make moral pronouncements. As the late Cornell University biologist William Keeton and his Princeton University colleague James Gould put it:

> ...science...cannot make value judgments: it cannot say, for example, that a painting or a sunset is beautiful. And science cannot make moral judgments: it cannot say that war is immoral. It cannot even say that a river should not be polluted. Science can, however, analyze responses to a painting; it can analyze the biological, social, and cultural implications of war; and it can demonstrate the consequences of pollution. It can, in short, try to predict what people will consider beautiful or moral, and it can provide them with information that may help them make value judgments about war or pollution. But the act of making judgments itself is not science. [From William T. Keeton and James Gould. 4th ed. *Biological Science*. New York, NY: Norton;1986.]

By proclaiming evolutionary theory and morality to be incompatible, creationists once again demonstrate a total misunderstanding of science and its limitations. Some have even blamed the teaching of evolution for a decline in moral values, citing this as a causative factor in the tragic 1999 Columbine High School shootings in Littleton, Colorado! Because creationists view the theory of evolution as contradicting their version of Biblical "truth" and because the Bible is their sole authority on both that "truth" and morality, they therefore view the teaching of evolutionary theory in public school biology classes as threatening "traditional values."

Another view, however, suggests precisely the reverse is the case in terms of a moral view of humanity in its proper relationship to nature. Ecologist Paul R. Ehrlich, Professor of Biology at Stanford University, has expressed this view as follows:

> Evolution is a necessary background for understanding anything about the real world, because our perceptual systems evolved. It is the way I make sense of biology at any level, and it surely a critical background for any ecologist. I believe the current ascendancy of creationism among the public in the United States (and, interestingly, *just* in the United States among western countries), is an important factor in worsening the human predicament. The decline in the teaching of evolution in schools cuts most Americans off from knowledge about their origins and the origins of the ecosystems in which they operate, and generally encourages a feeling of human exceptionalism. It has contributed to the misuse of both antibiotics and pesticides, resulting in the deaths of tens of thousands of human beings and increasing the odds of much more extensive disasters. Creationists are not, as many seem to believe, harmless—they help make the world a more dangerous place. [*A World of Wounds: Ecologists and the Human Dilemma*. No. 8 in the *Excellence in Ecology* Series. Oldendorf/Luhe, Germany: Ecology Institute; 1995]

It is often said that creationists simply do not understand evolution. As we have seen, this is certainly correct in terms of understanding the nature of its strong scientific underpinnings. Yet, as in the case of the Church with Galileo in the seventeenth century, creationists *do* understand the philosophical implications of evolution in terms of its view of the proper place of humankind within nature. They do not like that view at all. The Church-approved seventeenth-century view of humankind in the universe was a limited, self-centered one that placed us at its center. Being a human organization, the Church used the power derived from its close union with the state to lash out at those who, like Copernicus and Galileo, dared to burst this illusionary bubble. It was a similar human arrogance that was challenged once again by Charles Darwin. In this case, however, the paradigm shift moved humankind not just from the center of the universe but to being only one small part of the living world. In Ehrlich's words, it overthrew the concept of "human exceptionalism." As Harvard University paleontologist Stephen Jay Gould has put it, Darwin's work suggested that humans were merely "a fortuitous cosmic afterthought, a tiny twig on the enormously arborescent bush of life," rather than a species specially created and favored by God. It is perhaps not surprising, therefore, that Darwin's new paradigm, like that of Copernicus, unleashed a new storm of religious protest. In the case of Copernicus and Galileo, it became increasingly evident that science was right, the Church wrong, and Christian theology gradually adjusted to fit this new view. In the case of Darwin and the theory of evolution by natural selection, we live in an age analogous to the one that followed Galileo, in that there are still those who, like the creationists, wish to turn the theological clock backward rather than forward. Failing to convince the general population, like the medieval Church they turn to using the power of the state (for example, Kansas) to force their views upon society at large.

Although creationists may understand evolution and its philosophical limitations, they very distinctly do *not* understand science. As we noted earlier, its is not only biology but, in fact, the intellectual foundation of all the sciences that is threatened by creationism. The October 1999 issue of *Geotimes* was devoted to the Kansas School Board matter. In a series of articles appearing in this issue, several geologists and science educators reacted to the Board's decision. Linda Selvig, president of the National Earth and Space Science Teachers Association, expressed her view that:

> The proponents of removing evolution from the science classroom indicate that, because evolution is only a "theory" it holds no validity. Where they are wrong is that in the scientific community the term "theory" does *not* mean a "guess or hunch." Scientific theories are based upon a pre-

ponderance of evidence. The unique aspect of a scientific theory is that as new evidence emerges theories are modified or discarded. It is through knowledge of evolution and this questioning and process of self correction, for example, that we are able to openly discuss and understand the possibility of life on early Mars. Science is still looking for the "evidence for," not "proof of." [*Geotimes* 1999;44(October):21.]

One could hardly find a better statement pinpointing the precise manner in which creationists fail to comprehend the intellectual process we call science. They seem not to understand that someday a scientist might well come up with a better paradigm than natural selection for the driving force behind evolution or, for that matter, one to replace the concept of evolution itself. Instead, creationist literature has portrayed evolution as a godless theory dreamed up by atheistic scientists in order to lead students away from religion. During the Cold War between the United States and the former Soviet Union, creationist literature described evolutionary theory as a communist plot designed to accomplish precisely this end!

In fairness to religion, it should be pointed out that creationism represents an anti-intellectual phenomenon that is limited almost entirely to the continental United States. Even here, few if any modern-day Christian theologians find any difficulty in accepting evolution. Testimony in 1982 against the Arkansas law requiring the teaching of "creation science" was led by both scientists *and* theologians. In fact, many Christian scholars are embarrassed by creationism. In the same issue of *Geotimes* cited previously, The Reverend James W. Skehan, S.J., Professor of Geology and Geophysics at Boston College, referred to the Kansas School Board creationist vote as a "strategy of 'dumbing down' science teaching, not to mention beliefs fundamental to the Jewish and Christian religions." Nor is any conflict between evolution and religion to be found in Judaic theology, possibly because of its long tradition of respect for the intellectual process. This last point is especially significant, because, as noted, the Biblical passages cited by creationists as relating the "true" story of creation are found in Genesis, the first book of the Hebrew Bible. Finally, it should also be noted that Eastern religions, many far older than Christianity, express a viewpoint of humankind and its relationship with nature that is not only not in conflict with modern science but instead reasonably compatible with it. This is quite likely because, as Tibet's Dalai Lama stated in 1997: "If scientific research proved certain Buddhism teachings to be incorrect, I would agree that the teachings should be changed accordingly." Had similar attitudes prevailed earlier, the humiliation suffered by the both the Roman Catholic and Protestant Churches in the cases of Copernicus, Galileo, and Darwin, might well have been avoided.

Science and Religion: An Underlying Difference

One might well ask: Why have conflicts between science and religion, persisted since the rise of modern science in the seventeenth century? Almost certainly it is because, popular press accounts to the contrary, major and important differences separate the way the fields of science and religion view the universe.

In the early nineteenth century, Emperor Napoleon I of France met with the noted astronomer Pierre Simon de Laplace (1749–1827). Laplace was explaining to Napoleon his hypothetical model of the solar system incorporating the sun and the six then-known planets. The Emperor is alleged to have remarked: "But Monsieur Laplace, I see no mention of God in your model." La Place's reported reply was: "God? I have no *need* of that hypothesis!" The story is probably apocryphal but is illustrative of a major historical shift in philosophical thinking. Chemist G.N. Lewis (1875–1946) once noted that scientists "do not aim to seek [the] ultimate but [the] proximate [causes]." Ultimate causes are those that are concerned with "why" questions, whereas proximate causes are concerned with "what" or "how" questions. Explaining planetary motion or the circulation of the blood as the result of God's plan are examples of ultimate causes, whereas explaining the same phenomena by gravitational forces or the beating of the heart, respectively, are examples of proximate causes. Indeed, the rise of modern science in the seventeenth century may be said to have marked quite precisely a philosophical move from focusing on ultimate to proximate causes. This was *not* because of a rejection of a belief in God on the part of early scientists; on the contrary, many, if not most, were quite religious. Sir Isaac Newton, for example, saw the study of proximate causes as leading to revelations about the most ultimate of all causes, the mind of God. In brief, then, the rise of modern science marked a shift toward leaving the determination of ultimate causes to philosophers and theologians and concentrating instead on those proximate causes open to scientific investigation.

The continued conflict between science and religion rests on major philosophical differences between the two systems of thought. At various points in time, these systems may achieve a peaceful coexistence but only by ignoring their differences. Science is based on philosophical materialism, defined in Chapter 2 as the belief that all phenomena in the universe are the result of matter in motion, that is, the interaction of material entities in conjunction with known and measurable forms of energy (mechanical, chemical, electrical, etc.). As a philosophy, materialism admits no unknowable forces, no unmeasurable forms of energy, and no mystical or occult causes. Religion, on the other hand, is based on philosophical idealism, a view that stresses the primacy of ideas, nonmaterial entities, and abstract causes as significant causal agents. For example, in dealing with the origin of life itself, philosophical materialists would argue that matter has within its make-up the potential to form complex associations (for example, molecules), that these can form still more complex associations (membranes, cell organelles, etc.), and that the formation of higher level interactions does not need input from some mystical power. The known forces of nature—matter, energy, and their interactions—are all that is necessary to understand how life might have originated. Philosophical idealists would argue that something more than mere matter and energy is involved, whether that is expressed in a secular form as a "vital force" or in a religious form as the "will of God."

On the philosophical level, these are rather different ways of approaching an explanation of events in the world. Thus, a quiet truce has been established between science and religion by maintaining that the two approaches deal with different aspects of human experience. As just noted, Newton and others in the seventeenth century solved the dilemma by arguing that science deals with proximate and religion with ultimate causes. For many people, such a truce works in a practical, everyday sense. We may seek guidance and comfort in spiritual matters from literature, art, or religion, yet still turn to science or medicine to figure out how our bodies work and what makes a healthy diet. Many if not most people can accept a belief in a higher power in guiding their lives yet will not step out of a window because they have just prayed for gravity to be suspended. Thus the truce works, although it may often come at the expense of a rigorous philosophical consistency. For most people this appears as not too high a price to pay. It does emphasize, however, the underlying difference in philosophical outlook. This difference can and does surface on occasions when matters such as religious control of the school science curriculum emerge.

Conclusion

As we have just stressed, none of the preceding material about creationism and the triumph of materialist over idealist views on the origin of species should be interpreted as implying that religious beliefs should play no role in society. Religion has always been concerned with ethics and values, issues that every society must face. However, when we look to religious leaders, philosophers, or any others claiming expertise in ethical matters, it is imperative that they be scientifically informed about the natural world within which our moral and ethical choices are made. As philosopher Max Otto (1876–1968) noted in 1926:

> The universe is run by natural forces and laws, not by moral laws. However, human societies which live in the natural world must live by

moral laws. If those moral laws contradict or ignore the natural laws, it will be the human societies, not the physical universe, which suffer the consequences of such defiance. [From Max Otto. *Natural Law and Human Hopes.* New York, NY: H. Holt and Co.; 1926.]

There could be no better way to end this book than to underscore the importance of Max Otto's words quoted above. It is *precisely* those natural forces and laws with which modern science is concerned and to which, by its very nature, it is limited. Limited as it may be, however, science remains humankind's most powerful intellectual tool. As Otto noted, we ignore its findings at our peril.

Exercises for Critical Thinking

5.1. Distinguish between the "treasure-hunt:" and "social construction-ist" views of science. In what ways might you argue that both points of view are involved in understanding the factors influencing how science works?

5.2. How might social constructionist claims be tested? That is, even if such claims are true, how can those adhering to this view as a way of describing science test their claims against reality?

5.3. Political action groups supporting the movement known as "right to life" in favor of a constitutional amendment outlawing abortion, often use statements in defense of their position, such as "science has proved that life begins at conception." Analyze this statement in terms of the nature of science and its limitations.

5.4. Sociologist Troy Duster at the University of California, Berkeley, has argued that prenatal testing for supposed genetic defects, with an eye to abortion if the fetus exhibits any such defects, represents a "back door to eugenics." What might he mean by that phrase? Is such testing introducing eugenic practices back into society?

5.5. As more human clinical diseases become known to have significant genetic bases, drug companies and health maintenance organizations (HMOs) are requiring prenatal diagnosis before agreeing to cover a newborn on the parent's medical coverage plan. The HMOs' argument is that genetically determined traits are "prior conditions" and can be excluded on the same grounds that insurance can be excluded to someone who applies for a policy *after* having a heart attack. What are the ethics of such claims, and how should society respond to the accumulation of this sort of new knowledge?

Further Reading

Much has been written on all the areas of science and its social interactions as touched upon in this chapter. A few items for further exploration are listed here by chapter sub-headings

The social construction of science

Jones, Richard. The historiography of science: retrospect and future challenge. In Michael Shortland and Andrew Warwick, eds. *Teaching the History of Science*. Oxford, UK: Blackwell; 1989:80–99.

> *This is a good review of various methods of interpreting science over the past half century, from the "traditional approach" to social constructionist.*

Dear, Peter. Cultural history of science: an overview with reflections. *Science, Technology and Human Values* 1995;20:150–170.

> *A thorough discussion of what it means to interpret science in its cultural and social context.*

Flemin, Donald. Walter B. Cannon and homeostasis. *Social Research* 1984;51: 638–639.

> *A look at Cannon's work at the time of the Pareto seminars discussed in this chapter.*

Young, Robert. Malthus and the Evolutionists: the common context of biological and social theory. In *Darwin's Metaphor: Nature's Place in Victorian Culture*. Cambridge, UK: Cambridge University Press; 1985:23–55.

> *A classic essay showing the influence of Victorian social and economic metaphors on Darwin's thinking.*

The social responsibility of science

Allen, Garland E. Genetics, eugenics and the medicalization of social behavior: lessons from the past. *Endeavour* 1999;23:10–19.

> *A discussion of the development of eugenics and its political consequences.*

Paul, Diane. *Controlling Human Heredity*. Atlantic Highland, NJ: Humanities Press; 1995.

> *A very readable book that provides a good overview of the whole eugenics movement and its social consequences.*

Galston, Arthur. Education of a scientific innocent. *Natural History* 1971;80 (June-July):16–22.

> *Arthur Galston details his unwitting involvement in developing herbicides that wreaked havoc on the ecosystems of Southeast Asia during the Vietnam war.*

Jones, James H. *Bad Blood. The Tuskegee Syphilis Experiment.* (New York, NY: Free Press; 1981.
A detailed study of the Tuskegee experiment.

Scientific creationism and the relations between science and religion

Nelkin, Dorothy. *The Creation Controversy.* New York, NY: W.W. Norton; 1982.
This is a brief but useful summary of the creationist debate especially with respect to the Arkansas law of 1981.

Numbers, Ronald L. *The Creationists: The Evolution of Scientific Creationism.* Berkeley, CA: University of California Press; 1992.
Provides a valuable historical context for the development of various creationist movements in the United States over the past 150 years. The author places the modern creationist movement in a long line of evangelical tradition that has taken refuge in Biblical literalism.

Blanshard, Brand. *Reason and Belief.* London, UK: George Allen and Unwin, Ltd.; 1974.
The late author, a professor of philosophy at Yale University, includes a brief but excellent discussion of the case of Galileo vs. the Roman Catholic Church on pp. 58–62.

Suggested Websites

1. The Origin of Species by Charles Darwin

http://www.talkorigins.org/faqs/origin.html
What better source is there for the study of the science of evolution than Darwin's First Edition? Very readable despite the florid syntax of 19th century English, this electronic copy can be searched to discover Darwin's use of key words and phrases important to the history of evolutionary thought.

2. Image Archive of the American Eugenics Movement

http://vector.cshl.org/eugenics/
An archive of materials from the Eugenics Record Office at Cold Spring Harbor, which was the center of American eugenics research from 1910-1940, accompanied by essays from contemporary scholars.

3. Internet Resources on the Tuskeegee Study

http://www.gpc.peachnet.edu/~shale/humanities/composition/assignments/experiment/tuskegee.html
An extensive list of links and bibliographical references to this sad chapter in American medical history. Maintained by the Centers for Disease Control and Prevention.

4. Talk Origins Archive: Exploring the Evolution/Creation Controversy
http://www.talkorigins.org/

The mother of all sites dealing with creationist claims, "talk.origins is a Usenet newsgroup devoted to the discussion and debate of biological and physical origins. Most discussions in the newsgroup center on the creation/evolution controversy, but other topics of discussion include the origin of life, geology, biology, catastrophism, cosmology and theology. This archive is a collection of articles and essays, most of which have appeared in talk.origins at one time or another. The primary reason for this archive's existence is to provide mainstream scientific responses to the many frequently asked questions (FAQs) and frequently rebutted assertions that appear in talk.origins."

5. National Center for Science Education Defending the Teaching of Evolution in the Public Schools
http://www.ncseweb.org/

The NCSE is an organization working to "defend the teaching of evolution against sectarian attack" and a "clearinghouse for information and advice to keep evolution in the science classroom and 'scientific creationism' out." This site contains extensive links and up-to-date news on the struggle to keep creationism out of the public schools.

6. Dialogue on Science, Ethics, and Religion
http://www.aaas.org/spp/dspp/dbsr/about.htm

Initiated in 1995 by the National Science Foundation with three objectives: "(1) to promote knowledge about developments in science and technology within the religious community, (2) to provide opportunities for dialogue between members of the scientific and religious communities on significant topics for the sake of mutual understanding, and (3) to promote collaboration between members of the scientific and religious communities on projects that explore the ethical and religious implications of scientific developments."

Appendix 1

The Analysis and Interpretation of Data

For John Snow's analysis of cholera and more recently for researchers studying AIDS, observational information was critical in coming to conclusions about both the causes of disease and methods of transmission. Testing competing hypotheses requires the use of both observational and experimental data. Yet, a large collection of data is of limited value if it is not arranged in such a way as to reveal possible relationships. For example, John Snow's collection of mortality data on cholera in 1849 would not have yielded any significant conclusions had he not organized it by pinpointing the location of each death on a street map of London. This organization of data enabled him to see that the greatest number of deaths occurred in the vicinity of the Broad Street pump. How investigators choose to collect, organize, and display data determines to a large extent what information they believe is needed to answer the question at hand and what conclusions they wish to communicate.

Collecting Data and the Problem of Sampling Error

Many types of measurements in biology involve sampling small amounts of data from the vast collections potentially available. For instance, from a practical point of view it would be impossible to measure the height of all the individuals in a large city in order to determine the average height of the city's population. Not only would such an undertaking be time consuming and laborious, it would also be unnecessary, because by choosing a sample of individuals from the population at large an accurate set of data can be collected far more easily.

However, gathering that sample is not without potential problems. The most important problem is the potential for bias in the sample. If we sample 100 people from a city of 500,000, that sample must be representative of the population at large if it is to tell us anything meaningful about the population as a whole. If we sampled only people who happened to be basketball players, for example, the average height obviously would be considerably different than if we sampled people from a senior center. Thus, the sample would need to take into consideration all sorts of factors that might affect height: sex, age, ethnic background, etc.

Seeking Relationships: Collecting and Organizing Data

Data is collected, organized, and presented using many methods. Because these methods must be related to the question being asked, the nature of the phenomenon being studied, and the purpose of the investigation in the first place, researchers do not have one formula or set of rules for the process of collecting or organizing data.

Qualitative and Quantitative Data

Data collected directly from observations or experiments is called raw data. In Snow's study, for example, the raw data collected were the number of cholera fatalities as recorded hour by hour or day by day. Raw data may be collected and/or expressed in two forms: qualitative and quantitative. Qualitative data are those that are expressed in a general, non-numerical form. For example, the statement, "a lot of people died of cholera in London on September 3, 1849" is qualitative data, because it conveys simply the information that many people died. Qualitative data can be misleading, because by definition they are imprecise. For instance, what is "a lot" to some people may be only a few to others.

On the other hand, stating that "58 persons died in London on September 3, 1849" presents quantitative data. Quantitative data are expressed numerically and are therefore more precise. It is also possible for other investigators to test the accuracy of quantitative data more easily than qualitative data. Quantitative data also may be arranged into tables and charts, plotted into graphs, and subjected to statistical tests that tell something about their reliability or reveal relationships not otherwise apparent. Quantitative results from different experiments also may be compared more accurately than can qualitative data. In testing his water-borne hypothesis by comparing differences between the two water companies, for example, it was important for Snow to know how many people came down with cholera after drinking the water supplied by each company. If his results had stated merely that people supplied by Southwark and Vauxhall showed more cases of cholera than those supplied by Lambeth, it would be difficult to know if the difference was large enough to be significant.

By emphasizing the value of quantitative data, we are not suggesting that qualitative data are of no use in scientific research. For example, Snow found he could identify instantly the company from which each water sample came if he added silver nitrate to the sample, since it produced a milky precipitate, silver chloride, that could be identified immediately. However, Snow's method of distinguishing water from the Southwark & Vauxhall and Lambeth companies on the basis of cloudiness involved a qualitative judgment: How "cloudy" was enough to be considered positive? He could have made the test more quantitative by measuring the amount of silver chloride precipitate, but this was not necessary. His test clearly distinguished the source of the water, which was all he needed to know to test his hypothesis. Because qualitative data can often be collected more easily, it plays an important role in scientific research.

Measurement and Precision in Data Collection

Collecting quantitative data obviously involves some type of measurement. In Snow's case, the measurements with which he was dealing were counts of the number of fatalities resulting from cholera per 1,000 persons. Collecting such data is not as easy as it might seem. Snow had to be sure that the fatalities were the result of cholera because, even during the worst cholera epidemics, people died of other causes. If Snow had failed to separate out all other deaths, his data would have been worthless in determining the cause of cholera or its means of transmission. Attention to the detail of measurement procedures is absolutely necessary if measurements are to be of any value. If data are not obtained reliably in the first place, the most brilliant analysis may produce unreliable or meaningless conclusions. The computer science expression "garbage in, garbage out" applies to all scientific investigation.

Variations in Measurement by Different Observers

No two people see the same event or phenomenon in exactly the same way. Making a measurement always requires some human judgment and may be

Table A.1 Measurements of tail length in a sample of deer mice (*Peromyscus*)

Organism number	Observer 1	Observer 2	Observer 3
1	60.5 mm	60.2 mm	60.3 mm
2	61.0	59.9	61.1
3	62.2	62.0	63.0
4	68.1	68.0	67.9
5	60.7	60.6	60.2
6	58.3	58.4	58.5
7	66.6	66.7	66.3
8	56.7	56.6	56.5
9	62.5	62.6	62.5
10	60.8	50.9	60.5

prone to the introduction of small amounts of error. For example, in measuring a sample from a population of deer mice for tail length, three different biologists compiled the data shown in Table A.1 (only 10 organisms in the sample are shown here). In no case did all three observers get the exact same measurement from the same organism. Where, for example, does the tail actually start on a mouse? This means that if a team of investigators is working together making measurements, they must establish clear criteria among themselves for how to make those measurements. Establishing clear criteria increases the reliability of the data. Reliability simply means that the data are consistent and can be repeated by other observers. Reliability is a key feature of scientific data, because it means that other investigators can use the data without having to repeat all the measurements themselves.

Seeking Relationships: The Presentation of Data

A first step in the analysis of data is to arrange the data into one of several forms for inspection and display: distribution maps, tables, and graphs.

Displaying Data

Distribution Maps. Distribution maps show the localization of objects. For Snow, such a map showed the spread of cases of cholera over a given spatial area. Snow's distribution map of the location of households in which cholera fatalities had been reported made it possible for him to reveal clearly that the source of contamination was the Broad Street pump.

Tables. Tables represent another common way of displaying and analyzing data. A table consists of data arranged in two or more columns, enabling the viewer to see how items in one column relate to items in the other(s). For example, in following the course of the 1849 cholera epidemic, which began in late August and ran through mid-September, Snow collected the data shown in Table A.2. These data show clearly the quantitative progression of the epidemic.

One of the characteristics of any set of data such as those shown in Tables A.1 and A.2 is that they exhibit a range of values, from the highest to the lowest. For the data in Table A.2, the number of deaths during the period August 29 through September run from 1 to 145. The range therefore defines the outer limits of the measurements in any given sample.

Graphs. The data in Table A.2 also may be presented graphically. Graphs show the relationship between two or more factors arranged along two or more axes on each of which is plotted a specific scale of measurements. Figure A.1 shows a bar graph of the data in Table A.2. The horizontal, or x-axis, measures an independent variable, a variable not usually affected by the other factor or factors under consideration. The independent variable in Fig. A.1 is time, expressed in days of the month (which pass, after all,

Table A.2 Number of deaths resulting from cholera in the vicinity of Broad Street, South London, 1849

Date	Deaths resulting from cholera
August 29	2
30	10
31	58
September 1	45
2	120
3	57
4	50
5	35
6	20
7	29
8	15
9	14
10	8
11	8
12	2
Total	573

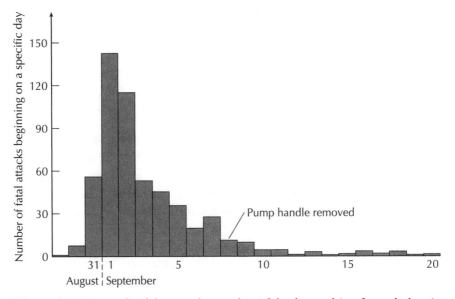

Figure A.1 Bar graph of data on the number of deaths resulting from cholera in London from August 29 and September 20, 1849. [From Snow (1855), summarized in Martin F. Goldstein and Inge F. Goldstein. *How Do We Know? An Exploration of the Scientific Proces*. New York, NY: Plenum Press, 1978: 40.]

regardless of whether cholera cases occur or not). The vertical, or y-axis, measures the dependent variable. As its name suggests, dependent variables are dependent on (that is, a function of) independent variables. The number of deaths resulting from cholera is a dependent variable, because it is a function of the time since the infectious agent arrived in the community. The point of intersection of the x- and y-axes is called the origin. The height of the bars in Figure A.1 illustrates the difference in number of cases from one day to the next. Bar graphs often make a quantitative point far more clearly and dramatically than a simple presentation of numbers in a table.

In a line graph, points representing the data collected are connected to one another by a line. Figure A.2 shows a line graph for the same data as that in the bar graph in Figure A.1. Each data point on a line graph relates a value on the x-axis to its corresponding value on the y-axis. The data points are plotted by finding the x-axis value (in this case, a specific day, for example, August 31) and then moving sufficiently high on the y-axis to find the y-axis value (in this case 58, the number of deaths occurring on that day).

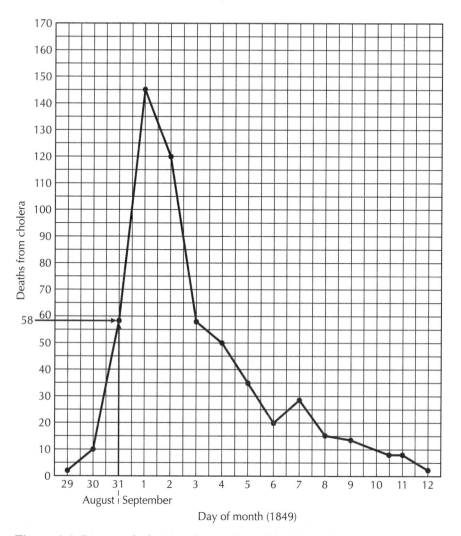

Figure A.2 Line graph showing the number of deaths resulting from cholera in London from August 29 to September 12, 1849 (same data as in Fig. 5.1). [From Snow (1855), summarized in Martin F. Goldstein and Inge F. Goldstein. *How Do We Know? An Exploration of the Scientific Proces.* New York, NY: Plenum Press, 1978: 40.]

Although both bar and line graphs show the change in death rate from one day to the next, in the bar graph the data for each day stand visibly separate from the others. The line graph, on the other hand, conveys a sense of continuity from one day to the next and thus emphasizes a trend over time. Both forms of presenting graphical material may be equally useful, depending on the phenomenon being studied and the point the investigator most wishes to emphasize.

Scales and Scalar Transformation

It is important to consider the scale on which the axes of graphs are arranged. Snow's data on number of deaths per day during the 1849 epidemic (Fig. A.1) show clearly the daily change, because the y-axis on which the number of deaths is plotted is laid out in units of 30. If the same scale had been laid out in units of 100 or 500, however, the graph would have looked quite different and shown the magnitude of the day-to-day changes less clearly.

The data in Figure A.1 are plotted on arithmetic scales, because both are based on numerically constant increments. Scientists sometimes use a logarithmic scale, in which each unit on the scale represents an increase by multiples, such as a two-fold or ten-fold increase: for example, 2, 4, 8, 16, 32, 64 . . ., or 1, 10, 100, 1000, 10,000, 100,000. A logarithmic scale is valuable in plotting data with a wide range of values or in which the rate of change in one factor is much greater than in the other. An example would be a graph plotting human population growth from prehistoric times to the present, with numbers ranging from 1 million to 5.5 billion. Because the numbers with which Snow was dealing ranged over a small scale (0–150 deaths over 15 days), an arithmetic scale was perfectly suitable. Scalar transformations, changing the scale of one or both axes of a graph, either from one arithmetic scale to another or from an arithmetic to a logarithmic scale, are often an important aid to analyzing data. They can also be used to convey very different messages (Fig. A.3).

Interpolation and Extrapolation

Scientists employ two techniques in constructing and interpreting graphs. Interpolation refers to the process of filling in, or generalizing, between two items of data in a table or graph. We employ interpolation whenever we draw a line on a graph connecting two data points. Suppose, for example, that Snow had recorded cholera deaths on an every-other-day basis (Fig. A.4). A solid line connecting data points for August 31 and September 2 would suggest that the number of deaths for September 1 should have been somewhere around 88–90. Thus, one function of drawing lines connecting data points on a graph is to interpolate predictions for cases not measured directly.

Interpolation may not always be accurate, however. Sometimes the discrepancy is trivial; at other times, it may be critical. As the dotted line in Fig. A.4 indicates, if Snow had based his analysis on data gathered every

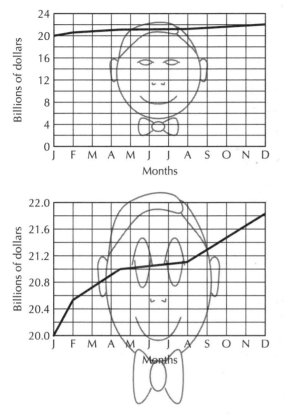

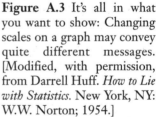

Figure A.3 It's all in what you want to show: Changing scales on a graph may convey quite different messages. [Modified, with permission, from Darrell Huff. *How to Lie with Statistics*. New York, NY: W.W. Norton; 1954.]

other day, he would have made two errors. First, he would have given the maximum number of deaths as 120 on September 1 rather than the actual 145. This could be a significant difference to public health officials trying to track the course of an epidemic. Second, he would have erroneously dated the peak of the epidemic as occurring on September 2 rather than September 1. This, too, might have been an important difference in later analyses of the time frame within which cholera epidemics develop. The interpolation in Figure A.4 gives the impression that the epidemic developed more slowly than was actually the case.

Extrapolation involves making predictions beyond the limits of the data available and is based on trends revealed by the data. Extrapolation of the data in Figures A.2 or A.4 involves predicting what the number of cholera cases might have been on September 13. The trend from September 6 or 8 to September 12 shows a steady decline. It would thus be reasonable to extrapolate that on September 13 there might be only one or

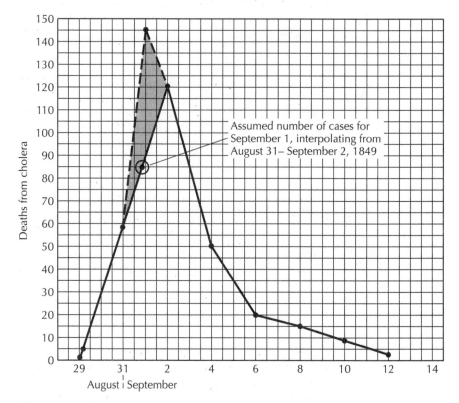

Figure A.4 Graph showing interpolation of data. Interpolation is an important part of generalizing any set of data, but it has the danger of misrepresentation. If only data on the number of deaths between August 31 and September 2 are taken into account, the interpolated value for September 1 would be 85 deaths. In reality, the number of deaths on that date was 145.

no deaths. Such an extrapolation would suggest that the epidemic was essentially over by September 13.

The Analysis and Interpretation of Data

Once a collection of data is organized into a table or plotted onto a graph, it can be more readily analyzed. Several common statistical approaches aid in the analysis and interpretation of data.

Correlation

In addition to showing the change of one quantity with respect to another, tables and graphs may also reveal a correlation. A correlation is a relationship between two factors, in which one factor changes in some way with respect to a change in the other. For example, from the map shown in Figure 3.3, Snow could have plotted a correlation between the distance people lived from the Broad Street pump and the number of cases of cholera. Such a plot is given in tabular form in Table A.3 and graphically in Figure A.5. Inspection of Table A.3 indicates that the number of deaths declines with distance from the pump. The exact nature of the decline, that is, the degree of correlation, is not clear from the numbers alone, however. The graph in Figure A.5 shows the relationship more clearly. The fact that the points of data form a fairly a straight line suggests that the correlation between the two factors is a strong one. In this case, the correlation is said to be negative, in that as the numerical values for distance (the independent variable) increase, the numerical values for number of deaths (the dependent variable) decrease. If the number of deaths from cholera increased as distance from the pump increased, the correlation would be said to be positive.

The slope of a graph line also tells us the precise quantitative way in which the two factors correlate with one another. When the slope is at a 45° angle, as shown in Figure A.5, the correlation is said to be 1:1; that is, as one factor increases by a specific unit, the other factor changes by a comparable unit. For example, with every 50-yard increase in distance from the pump, the number of deaths decreases by about 15. The lines on a graph of correlation, however, may be of any slope. As long as the units are the

Table A.3 Snow's data on distance from the Broad Street pump and dates and numbers of deaths resulting from cholera

Distance from Broad Street pump (in yards)	Deaths reported resulting from cholera (Aug. 29–Sept. 12, 1849)
0–50	130
100–150	100
150–200	80
200–250	55
250–300	45
300–350	30
350–400	15
400–450	1
450–500	0
Total	573

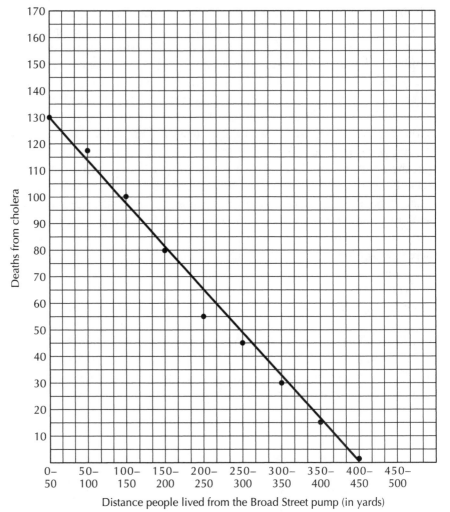

Figure A.5 Graph showing correlation between distance people lived from the Broad Street pump and deaths resulting from cholera in 1849. In this case, the correlation is said to be a negative one, because as one factor, the independent variable (distance from the pump), increased the other factor, the dependent variable (deaths from cholera) decreased.

same from one graph to another, a slope steeper than 45° means that as the independent variable changes by one unit the dependent variable changes by more than one unit. Conversely, a slope of less than 45° means that as the independent variable changes by one unit the dependent variable changes by less than one unit.

It should be apparent that graphical representations of correlations may be compared with one another only if they are plotted in comparable units. For example, we might alter the slope of the line simply by choosing a different scale on which to plot the number of deaths but leave the distance at the same scale. If, instead of plotting deaths in units of tens (10, 20, 30, 40 . . .) as shown in Figure A.5, we were to plot it in terms of twenties (20, 40, 60 . . .) as shown in Figure A.6, the slope of the line would be much less steep but the correlation itself would not be changed. What *would* be changed would be the perception of the viewer.

Other than by visual inspection of the distributions of points on a graph, how can we determine whether a particular correlation is strong or weak? To get an estimate of how closely the trend in one factor is associated with that in the other factor, statisticians calculate a value known as the correlation coefficient. A correlation coefficient expresses the ratio of change in the two factors being compared in any given data set. Correlation coefficients run from -1 to 0, and from 0 to +1, with values on the minus side representing negative correlations and those on the plus side representing positive correlations. A correlation coefficient of 0 indicates no relationship between the two entities being measured. For example, using Snow's data correlating the number of cases of cholera with the distance

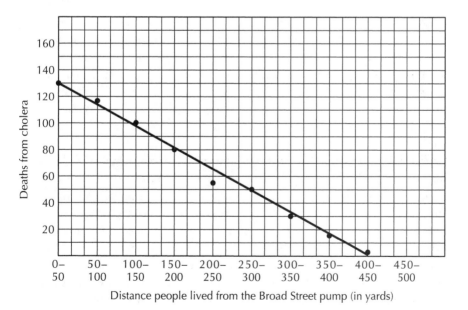

Figure A.6 Same data seen in Figure 5.5 but with data on the y-axis plotted in intervals of 20 rather than intervals of 10.

people lived from the Broad Street pump, we get a correlation coefficient of -0.8. This represents a strong negative correlation: that is, the greater the distance the fewer the cases. We could also calculate the correlation as a positive one if we compared proximity to the pump.

The method of calculating correlation coefficients is sufficiently complex that it will not be included here, but knowing the correlation coefficient gives an immediate, and quantitative indication of the degree of association between the entities being compared. Correlation coefficients also make it possible to compare sets of data that may be based on different units or types of measurements. For example, we could still compare results if other studies of the Broad Street pump had measured distance in meters rather than yards or in amount of water consumed from the pump by individuals rather than distance they lived from it

It must be stressed that *correlations do not by themselves tell us anything about cause and effect.* For example, the data given in Table A.3 and plotted in Figure A.5 do not necessarily tell us that the water from the Broad Street Pump was the actual cause of the deaths from cholera. Recall that Snow had difficulty convincing people of the correctness of his water-borne hypothesis. Proponents of the effluvia hypothesis could always argue that, because the pump was a center of activity and the area around it was usually crowded, effluvia from infected individuals meeting at the pump, rather than water from the pump itself, was the cause of infection. Thus the existence of a correlation alone is not enough to establish a causal agent. Snow needed, in addition, the data obtained by comparing the populations using water from the two different water companies to demonstrate a strong cause-and-effect connection between the source of water and occurrence of cholera. By showing that only people who drank water from the Southwark & Vauxhall Company contracted cholera, Snow was able to establish a plausible connection between polluted water and the spread of cholera.

Correlations may be quite seductive, suggesting what looks like an obvious cause-and-effect relationship but is completely spurious (for example, the almost perfect 1:1 correlation between the increase in a person's age from 1950 to 1980 and increasing levels of pollution in major urban areas). In other cases, however, suggested correlations might well turn out to be accurate. Indeed, one of the great benefits of establishing correlations in the first place is that they suggest *possible* causal relationships. However, an actual causal relationship can only be established by doing further research and gathering additional data on the system or organism in question.

Rate and Change of Rate

When John Snow calculated deaths in the 1853 London cholera epidemic he presented his data not only as the absolute number of deaths (Table 5.2) but also as deaths per number of households, that is, as a death rate. Rate is a measure of change in quantity of the dependent variable per some standard unit of the independent variable, such as time, volume, etc. Snow's death rate measured number of deaths per standard unit of population, in this case per 10,000 households. Calculating rate makes it possible to compare different samples using the same common denominator; in Snow's case a per capita basis or standard unit of population. For example, comparing the total number of deaths for houses supplied by each of the two London water companies would have been meaningless if given only in absolute numbers, because one company might have served a larger number of households than the other. As shown in Table A.3, by using death rate, Snow was able to demonstrate clearly that one company was associated with a far greater per capita incidence of cholera infections than the other.

Analysis of Distributions: Central Tendency and Dispersion

In trying to determine patterns of infection in cholera patients, Snow and others encountered considerable variation in the time required for the disease to run its course, recorded as the time required for a patient, once infected, to either die or recover from the effects of the disease. For some individuals, the time between infection and death or recuperation was very short: 1 or 2 days. In others, it took up to a week. The average, or mean, was around 3 to 4 days. If Snow had wanted to represent this variation quantitatively, he might have tabulated the results from a number of individual cases and plotted them on a distribution graph (Fig. A.7). Distribution graphs plot a range of measurements on the x-axis against the number or frequency of individuals in any particular category on the y-axis. These graphs are generally used to show the distribution of measurements around a mean or average for a population of individual organisms, for example, height in a group of people (Fig. A.8).

Two very useful means for describing characteristics of a data set of such measurements are the central tendency and the dispersion of the data. Central tendency is given by three values: the mean, median, and mode. The mean is the average value for a group of measurements and is calculated as:

$$\text{Mean} = x = \frac{\sum x_i}{n}$$

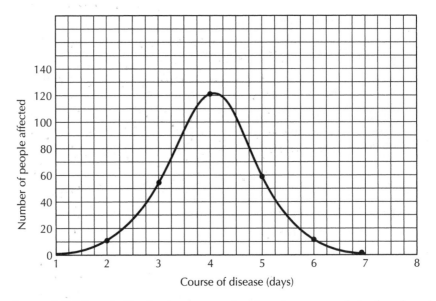

Figure A.7 Normal distribution curve, showing frequency of deaths plotted against number of days since the epidemic began. The graph shows that the disease peaked around the fourth day after the first infection was noted.

where x_i are the individual measurements, n is the total number of measurements, Σ means "sum of," and x is the symbol for mean. The mode is the most common measurement in the sample, and the median is the value above and below which lie equal numbers of measurements. Measurements of central tendency often give a bell-shaped curve, or curve of "normal distribution," as shown in Figures A.7 and A.8. Early statisticians were fascinated to discover that measurements of an extremely wide variety of samples showed some variant of the normal distribution curve.

What the measures of central tendency do *not* show, however, is the dispersion, that is, how widely or narrowly the sample is dispersed around the mean. Variance and standard deviation are both measures of dispersion, that is, how much the sample as a whole deviates from the mean. In calculating variance we cannot simply average the deviations from the mean, because the negative deviations will be cancelled out by the positive (in a normal curve) and the result will be zero, not a very useful number! It is possible, however, to average the squares of the deviations, using the following calculation:

$$\text{Variance} = \frac{\Sigma (x_i - x)^2}{n - 1}$$

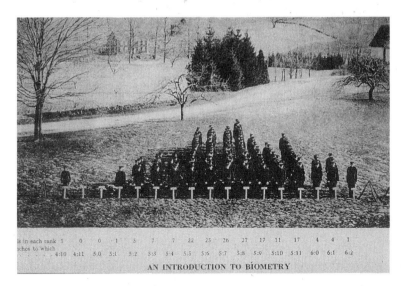

Figure A.8 A famous photograph from a 1914 textbook illustrating the normal curve of distribution for height in a group of 175 World War I recruits. Each small cross in front of a row of individuals indicates one particular height category (such as 5'8" or 5'9", as indicated by the numbers below the photograph). [Reprinted from A.F. Blakeslee, *Journal of Heredity* 1914:5.]

Squares of the deviations are divided by $n - 1$, because dividing by n alone tends to underestimate the variance, especially if the population is relatively small. Variance is a useful measure, but it is more common to estimate dispersion by calculating the square root of the variance, or what is called standard deviation, as follows:

$$\text{Standard deviation} = \sqrt{\frac{\sum (x_i - x)}{n - 1}}$$

One major reason for using standard deviation is that the units for variance itself (for example, height squared) often do not make sense (what, after all, is squared height?). However, by taking the square root of the variance, we come out with real units of measure (height in inches, number of days for individuals to show symptoms of cholera, etc.). Taking the square root compensates for squaring the measurements in the first place. The value of standard deviation is that it tells us a great deal about dispersion of the data around the mean. For a normal distribution as shown in Figures A.7 and A.8, 68% of all the observations lie within 1 standard deviation on either

Figure A.9 Normal distribution curve showing standard deviation limits. Standard deviation provides a basis for comparing distributions of real data against the expectations of where measurement values might ordinarily be expected. Thus, in a mathematically perfect distribution curve, 68% of the measurements will fall within 1 standard deviation from the mean, 95% will fall within 2 standard deviations of the mean, and 99% will fall within 3 standard deviations.

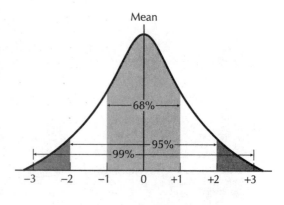

side of the mean, and 95% lie within 2 standard deviations on either side of the mean. Ninety-nine percent of the measurements lie within 3 standard deviations (Fig. A.9). In calculating standard deviation(s) for any set of data, the higher the value, the greater the dispersion of the data around the mean and thus the greater the width of the bell curve.

All distributions do not follow a normal curve, however. Some distributions may be either bimodal or skewed. A bimodal curve (Fig. A.10A) is one in which there are two modal groups, indicating at least two major groupings of characteristics within the sample. Field naturalists often find bimodal distributions within a particular population of organisms in nature, for example, a black and brown form of ground squirrels or left- and right-coiling shells in snails from the same population. The two modal groups do not have to be equal in frequency for the distribution to be described as bimodal. However, they do have to be distinctly demarcated from each other such that each has its own mean value. Skewed distributions are those in which the mode is distinctly different from the mean, so that the peak of the curve is not symmetrically placed between the two extremes of the range (Fig. A.10, B). The mode may be skewed to the right above the mean or to the left below the mean. Skewed distributions simply indicate that many individual measurements in the sample lie toward one side or the other of the mean, not equally in both directions.

Levels of Significance

The question of significance deals with determining whether given measurements represent a meaningful or chance departure from the expected.

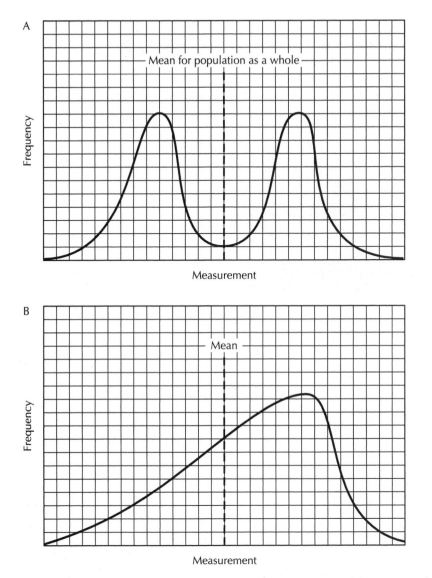

Figure A.10 A. Bimodal distribution graph showing the two modal groups, with the mean in the "valley" between. B. Distribution skewed to one end of measurement range (in this case the high end).

Suppose, for example, we measure height in a group of 100 college students on a campus in Minnesota, and get a normal curve with a mean of 5'8". Suppose that we then get another much smaller sample of 10 students from a college in California and find that the mean height is 6'4". The difference might reflect actual differences in the populations in the two localities, or it might reflect a bias resulting from the much smaller size of the second sample. The question investigators in this sort of situation would like to know is whether this measured difference is significant or not; or, expressed another way, what is the probability that the difference we observe in the small sample from California occurred purely by chance?

Starting with the null hypothesis that there should be no significant difference between the two populations, we then try to reject that hypothesis. If the probability of making a certain measurement or observation is less than 5% (that is, would occur in less than 5 instances out of 100), the results are considered to be significant and the null hypothesis can be rejected. Investigators often set their standard of significance level before making their measurements. They decide ahead of time what the probability would be of obtaining, by chance alone, a difference as large or larger than the observed difference. For example, knowing that our second measurement of the California college students only consisted of 10 individuals, we could calculate by statistical methods what the chances are that we would get such a different mean by chance alone. A significance level of 0.05 (or 5/100) is the generally agreed-upon standard for stating that the results are not purely a product of chance and therefore that the difference is a significant one. Investigators who wish to be more rigorous may analyze data to the 0.01 (1/00) significance level. These levels are somewhat arbitrary, but only in the sense that the values of 0.05 and 0.01 represent two agreed-upon standards for judging how likely it would be to get a significant difference in two sets of data by chance alone.

Further Reading

Barnard, Chris, Francis Gilbert, and Peter McGregor. *Asking Questions in Biology*. New York, NY: John Wiley; 1993.
  A well-written and clear presentation of many aspects of hypothesis formulation, data analysis, and experimental design. Aimed at the undergraduate level, contains many good examples, especially from the field of animal behavior.

Gilbert, Neil. *Biometrical Interpretation: Making Sense of Statistics in Biology*. Oxford, UK: Oxford University Press;1989.
  This is a concise, well-written and user-friendly introduction to a variety of statistical concepts, all using biological examples. Explanations are in simple language, and the book overall requires minimum mathematical background.

Huff, Darrell. *How to Lie with Statistics* (New York, NY: W.W. Norton; 1954). *Although almost 50 years old, this is a clever, simple, and very well-written book that provides a good introduction to statistical thinking and especially to methods of representing data. The author's sense of humor and the clear illustrations make the subject of statistics not only interesting but painless.*

Exercises for Critical Thinking: Possible Answers

Chapter 2

2.1 *Possible answer:* An observation is a discrete item of sensory data (for example, "the frog jumped"). A fact is an observation or set of observations that are agreed upon by a group of people. A conceptualization is a more general or abstract statement that goes beyond concrete facts and relates the facts to each other.

2.2 *Possible answers:*
   a. Fact. An agreed-upon set of observations.
   b. Observation. A specific item of sense data, visual in nature.
   c. Conceptualization. The statement offers an abstract reason for why the planets move in the direction they are observed to move.
   d. Conceptualization (same reason as c)
   e. Conceptualization. The statement is a generalization going beyond the specific set of green apples the observer has tasted.
   f. Observation. The statement is about the contents of the report derived from reading the document itself.

2.3 *Possible answers:*
   a. (i) At dusk, the light is dim but not perceived as quite as dim as it might actually be, so a person's vision is more impaired than they think. This hypothesis could be tested by evaluating a person's vision under various levels of illumination and also by gathering more precise data on degree of illumination compared with the accident rate as afternoon wanes into dusk. (ii) An alternative is that dusk is also "rush hour" on certain days anyway, so traffic is higher and accidents are more frequent. This hypothesis could be tested by measuring accident rate (number of accidents per total volume of traffic) for different periods during evening rush hour.

b. When the tumbler is warm, the air inside is expanding, forcing the bubbles to the outside. As the glass cools on the drainboard, the air pressure outside becomes greater than that inside, so the bubbles are forced back inside. A test would be to wash the tumbler in cold soapy water, in which case it would be predicted that no bubbles would form. A student once suggested that there might be a fan on in the room at the time and that this created air currents that caused the air to exit the glass; cooling of the glass then caused the air to reenter. What might be some problems with this explanation? Could the explanation be tested?

c. The results suggest that contact between offspring and the strain A mother is essential for cancer to develop. A likely transmitter at this stage of development is the mother's milk, which might contain a virus that inserted itself into cells of the young mice and later caused those cells to become cancerous. Several varieties of cancer are known to be associated with viruses or genes of viral origin that have inserted themselves into the cells of their host.

2.4 *Possible answers:*

a. *If* diminishing light intensity causes cricket chirping to slow down, *then* crickets placed in the laboratory when the light was successively diminished, should show decreased chirping

b. The results contradict the hypothesis/prediction. The slight variation in chirps at different intensities shows no trends.

c. *If* diminishing temperature causes cricket chirping to slow down, *then* at successively lower temperatures crickets should show decreased chirping.

d. The results support the second hypothesis, because the rate of chirping decreases markedly with decrease in temperature. The trend here is very clear.

2.5 *Possible answers:*

a. Valid reasoning; a true conclusion deriving from two false hypotheses

b. Valid reasoning; a true conclusion deriving from two true hypotheses

c. Valid reasoning; a false conclusion deriving from two false hypotheses

d. Invalid reasoning; a false conclusion from two true hypotheses

e. Invalid reasoning; a false conclusion from two false conclusions

2.6 *Possible answer:* Basic feelings encountered when an old paradigm begins to provide problems or anomalies are confusion and sometimes resentment or anger. Generally old paradigms are not abandoned until new ones are found to replace them. The acceptance of a new paradigm is often accompanied by a sense of great discovery, illumination, excitement, and relief.

2.7 *Possible answers:*

a. It would be reasonable to hypothesize that if allowed contact with her kid for even 5 minutes a doe will establish a bond, perhaps based on sight or smell recognition cues. She will recognize the kid on return even after an hour or more. However, if she is not allowed to establish any sight-smell bond, she will reject the kid as foreign. Once the bond has been established in the first 5 minutes, the doe is distressed at the kid's removal, but, if no bond is ever established, the doe's maternal behavior is never evoked and she behaves as if she never gave birth. This hypothesis could be tested by substituting a foreign kid for the doe's own natural kid immediately after birth, then removing it, and returning it an hour later. If the doe accepted the kid this would reinforce the smell-sight bond hypothesis. If she rejected the foreign kid an hour later this would tend to refute the smell-sight bond hypothesis.

b. The second set of observations/experiments confirm the smell-sight bonding hypothesis, because the doe accepts even a foreign kid if she is allowed 5 minutes with it after birth.

c. It would be possible to distinguish between sight and smell as the possible avenues of recognition used by the doe. The doe's nostrils could be plugged with scented cotton that would block her sense of smell and the two experiments above (using her own kid and then a foreign kid) repeated. The converse could be done by covering the doe's eyes. A third approach would be to rub a foreign kid with liquids from the placenta of the doe's natural kid, allow the doe to spend 5 minutes with her own kid and then remove it. The foreign kid, now smelling like the original kid, could be returned 1 hour later, with the following predictions as to the doe's behavior: *If* smell were the main means of recognition, *then* the doe ought to accept the foreign kid; *if* smell is not involved or *if* sight is the main means of recognition, *then* the doe ought to reject the foreign kid.

2.8 *Possible answer:* The answer to this problem is not intuitively obvious. You can start by ruling out the tenth draw: If you draw eight white marbles in a row, there can only be two marbles left, both of which must be red. This leaves only draws one through nine as possibilities.

It might seem reasonable to start with the ratio of white to red marbles: If there are eight white marbles and ten red marbles, then that is a 4:1 ratio of red to white marbles and therefore the chances of getting a red marble on the first draw are one out of five, or 20 percent. This is most certainly correct. If you do get a white marble on the first draw, that leaves only seven white marbles in the bag to two red marbles and therefore the odds of getting a red marble on the second draw increases. Still more of an increase in the odds of getting a red marble on the third draw occurs if a white marble is drawn on the second draw, and so on. Generally overlooked, however, is the fact that, since the odds of getting a red marble on the first draw are 20 percent, this means that 20 percent of the time one never gets to the second draw! If we enter that factor into the equation, it becomes quickly apparent that the first draw is the one most likely to get a red marble, the second the next most likely, the third the next most likely, with the odds decreasing down to the ninth draw, the least likely possibility.

Chapter 3

3.1 *Possible answer:* Hypotheses that cannot be tested, however interesting or imaginative (or farfetched) they may be, can never be either supported or rejected and thus add nothing in the long run to our understanding of the world. Formulating hypotheses without the check of testability also gives free reign to nonrigorous, sloppy thinking. It is possible to propose anything if you are not under any obligation to test it in the real world. On the other hand, many hypotheses that seemed untestable when first formulated turned out to be imaginative enough to stimulate later experimentation. So, lack of immediate testability is not always a reason for dismissing a new hypothesis. For example, many of the early theories about the genetic code were purely theoretical hypotheses that no one could immediately test, but they stimulated thinking about the code as a "language" and eventually led to very fruitful predictions and biochemical tests.

3.2 *Possible answer:* Hypotheses about phylogenetic relationships or past geological occurrences can only be tested by observation of fossils or geological strata. Other examples include (when human subjects are involved) the hypothesis that smoking causes cancer, that Amerindians descended from Asian migrants who crossed the Behring Strait at some point in the past, that teenagers present a greater driving risk than those who are 25 years old, that women inherently like child care more than do men, and others. Hypotheses testable by experimentation are preferred, because the variables that might influence the outcome can be controlled and the experimenter can set up conditions to his or her specifications. Experiments allow a more rigorous testing of alternative hypotheses by making predictions and then manipulating the system so as to observe whether the outcome does or does not verify the prediction.

3.3 *Possible answer:* Many problems are associated with this study. (A) First and perhaps most obvious, the average age of the two populations is not the same, the reformatory group being on average a year older than the Latin school population. In adults, a 1-year discrepancy in age might not be all that serious. However, during adolescence, when 1 year can make a huge difference in growth, especially where the characteristic being measured is body form, the discrepancy can be significant. The two populations also are very likely not matched for socioeconomic factors, which can have important influences on body growth and rate of development. It is also not clear whether ethnic differences were taken into account, differences that clearly might have some relation to body shape. Definition of body forms is at best a vague and largely subjective process. Look around at your friends and try to make such classifications. One more problem is that the underlying assumption that body shape somehow determines behavioral tendencies. Correlation of a physical trait with a behavior does not establish cause and effect. (B) The study raises some ethical problems which, if its conclusions were widely accepted (which they were for a time in the 1940s and 1950s), could lead to prejudging some boys by superficial examination of their appearance, and toward expectations that they would have a greater tendency to delinquent behavior. Such expectations often elicit the predicted results, becoming "self-fulfilling prophecies." No indication was given that the boys in either group were asked to sign consent forms to have their physiques measured. This aspect of the issue emerged several years ago when a woman opened a textbook and found a picture of herself naked. The photograph came from a body-form study that had been a required part of the physical education program at her college in the 1950s. She was rightfully angered at what she considered a basic invasion of privacy.

3.4 *Possible answer:* Diseases are caused by infective agents, such as bacteria, viruses, parasitic microorganisms (among others), and environmental toxins of various sorts. Understanding how these affect the human body is obviously one of the aspects of fighting disease. On the other hand, disease is spread by interaction between people, which includes a wide variety of social factors from individual and group behavioral practices (types of interpersonal contacts), as well as more widely organized applications, such as public health practices, such as how are wastes in a community disposed of, how are the rights of the individual to be balanced against the overall public health interests of the community?.

3.5 *Possible answer:* This is always a delicate balance. Quarantine has been used during many epidemics, but the extent to which it is employed needs to take into account: (A) How easily transmissible is the disease? (B) What is the form of transmission (sexually, through water supply, soiled clothing, etc)? and (C) What community mores and standards affect these issues? Restricting entry (immigration) to an area can be quite difficult, especially when a community gives way to panic about an epidemic or possibility of an epidemic. At the height of the AIDS epidemic in the United States (early 1990s), there was a call for the U.S. Immigration and Naturalization Service to bar admission to all immigrants who were HIV positive. Such a prohibition never became actual law, but there were attempts at many ports of entry to find out which prospective entrants might be infected.

Chapter 4

4.1 *Possible answer:* Highly subjective judgments are involved here, of course. The authors would rank (c) first, (a) second, and (b) third. The extinction of the dinosaurs had for many years been hypothesized as the result of climate change, with the resulting disappearance of the plants upon which the herbivorous dinosaurs fed (and thus, in turn, the loss of these dinosaurs as food for the carnivorous forms to prey upon). The Alvarez hypothesis, with its associated meteor impact crater prediction, presented a whole new way of looking at things and suggested entirely new avenues of research. The nerve growth factor research was certainly vitally important in the field of developmental biology and also influenced the direction of research in the field but did not represent as dramatic a shift in this direction. Hasler's salmon migration study was a monumental contribution to the field but only in providing experimental evidence for what had been suspected for some time—that the chemical composition of salmon home streams provided the clues enabling them to return to their home streams to spawn. In none of the three cases, however, is the level of magnitude of paradigm shift even close to that of Dalton's atomic theory in chemistry, quantum mechanics in physics, or the Darwin–Wallace theory of evolution by natural selection in biology

4.2 *Possible answer:* The findings suggest further support for the Darwin–Wallace paradigm. That such widely differing organisms possess compatibility at the tissue-cell-molecular level would be predicted by the initial hypothesis of evolution by natural selection, which, of course, suggests an ancestral relationship between all species.

4.3 *Possible answer:* Because the Alvarez K–T boundary hypothesis associates a high level of iridium with meteoritic impact and suggests that event as the causative factor in mass extinction, the new discovery amounts to a false prediction of the Alvarez K–T hypothesis and thus that hypothesis itself must be false. Being human, however, scientists tend to be very fond of their hypotheses. Thus, for example, in defending their hypothesis, the Alvarez's might suggest that the fossil record during the reported time period was not complete enough to record a statistically significant level of extinctions and that the meteoritic crater, like the one eventually found at the Yucatán Peninsula, had simply not yet been located.

4.4 *Possible answer:* The nerve growth factor (NGF) and salmon homing cases involved the greatest amount of experimentation. In the case of NGF, observation was involved, not so much in testing but in formulating a hypothesis. The observation that mouse sarcoma greatly stimulated neuronal growth gave rise to the hypothesis that the sarcoma was producing a substance that directly affected neuron growth and maintenance. Similarly, Hasler's work on homing in salmon began with the observation that salmon return to spawn in the very streams in which they were originally hatched. The Nemesis case is the one in which observation was used most regularly to test aspects of the hypothesis of meteor impact: for example, searching for periodicity in the paleontological record or for remnants of an impact crater that would match the estimates of the meteor hypothesized to have struck the earth 65 million years ago.

Chapter 5

5.1 *Possible answer:* The "treasure hunt" concept of science assumes real scientific laws in nature that have an existence independent of time and place. The scientist's job is to use clues to find the treasure, which will be the same no matter who discovers it or when. The social constructionist view is that scientists "construct" a view of nature using the tools of language, metaphors, philosophy, and analogies available to them, and that because these tools change from one culture to another, the resulting view of nature necessarily will reflect time and place. It could be argued that both points of view are important in assessing how science is pursued. There can be little doubt that our language, metaphors, comparisons, and analogies play an important part in constructing and communicating to others our view of nature. However, if we assume a real world out there beyond our senses, then our socially constructed view of that world still will have to be tested in reality. In that way, different constructions can be compared and contrasted and the most fruitful ones chosen to develop.

5.2 *Possible answer:* Social constructionist and other such views cannot be tested directly, of course. For example, we can put forth as a scientific hypothesis that both Darwin and Wallace were influenced by the social, political, and economic environment of nineteenth-century Great Britain. This hypothesis would predict that, had these two men lived in a socialist or a precapitalist society, their paradigm would have been expressed using different metaphors, emphasizing perhaps cooperative rather than competitive aspects of nature. Quite obviously, such an experiment cannot be performed. However, it *is* possible to make comparisons between hypotheses devised in different cultures and, in that way, gain some insight into how social and cultural factors affect the way science is done. Although such comparisons are subject to other interpretations, they provide one way to test predictions from the social constructionist perspective.

5.3 *Possible answer:* As emphasized in Chapter 2, science cannot *prove* anything. It can only establish its "truths" in terms of probabilities. The statement contains other errors as well. First, even if a "moment" were a precisely defined unit of time (as is a millisecond, for example), there is no one "moment of conception." Fertilization of an egg by the sperm is a process, not an instantaneous event. Considerable time elapses between initial contact of sperm with the egg surface and the fusion of the male and female pronuclei and still more before the initiation of their combined genetic underpinnings of development. Even more time will elapse before it is determined if the resulting fertilized egg (zygote) will finish its developmental journey down the Fallopian tube, implant in the uterine lining, and initiate pregnancy, or, as appears to be the case with as many as half of successful fertilizations, be aborted naturally and pass unnoticed through the vagina. Second, the statement also implies that "life" is a clearly defined entity. It is not. One has only to pick up a reasonably decent high school biology textbook to learn that, as early as the nineteenth century, biologists recognized that attempts to define life were fruitless. Claude Bernard (1813–1878), considered by many to be the father of modern physiology, noted:

> ...it is necessary for us to know that it is illusory and chimerical and contrary to the very spirit of science to seek an absolute definition of [life]. We ought to concern ourselves only with establishing its characteristics....

In fact, all science can do is attempt to describe those characteristics displayed by most (although by no means all) life forms. This being the case, therefore, we can state with certainty that life most definitely does *not* begin at any "moment of conception," because clearly both egg and sperm are alive, as are their progenitor cells, their progenitor cells, and so on, back perhaps 4.5 billion years ago to the origin of forms meeting some threshold number of features characterizing living matter.

We must stress here that we are not suggesting that one may not oppose abortion for any variety of reasons including, as is often the case, the religious conviction that it is "immoral," but only that one cannot use science, most especially the mistaken concept of scientific "proof," to support that position. As the quote from Keeton and Gould cited in this chapter makes clear, science has nothing to say on the morality of such issues as abortion.

5.4 *Possible answer:* Much depends on how eugenics is defined. The older, historical meaning was tied to state-sponsored programs and had a coercive quality about it. The older meaning of eugenics also was tied to concerted efforts to "improve the race" through planned breeding. Today, more subtle forms of coercion, such as denial of health care coverage, could end up forcing families to make choices not unlike those they would have been forced to make under older eugenic legislation. At the same time, the modern movement is not so motivated by overt claims to "improve the human species."

5.5 *Possible answer:* It could be argued that life in general is a "genetic disease," because we are all going to die of something at some point. This argument would hold that maximizing our potential in the time we do have is more important than figuring out cost-benefit analyses of human worth. On the other hand, one might argue that genetic diseases often are very expensive to treat and, even if the treatment allows the individual to contribute something to society, in the balance it is an inefficient way to manage limited health care resources. Where diseases—genetic diseases in this case—can be prevented, they should be. Cures are only for those diseases that cannot be prevented in the first place.

Index

A

Abiogenesis, 25–28
Adenosine triphosphate (ATP), 151–152
African Americans, in Tuskegee Study, 164–168
Agent Orange, 162–163
Agent White, 162–163
AIDS epidemic, 11, 91, 101–111
Allamandola, Louis, 27–28
Allen, G. E., 172
Alvarez, Luis, 138–140
Alvarez, Walter, 138–140, 144
Anomalies, 67
Applied research, 151
Archaeopteryx fossil, 173
Arithmetic scales, 194
Arrhenius, Svante, 25
Astrology, 168
Atomic bomb, 149
Average, 201
Axons, 116–122

B

Baker, J. J. W., 172
Baltimore, David, 71
Bar graphs, 193
Bernard, Claude, 2–3, 155, 158

Bernstein, Max, 27–28
Bias in science, 57–66
    conscious, 58–59
    fraud and, 57–59
    unconscious, 59–66
Big bang theory, 169
Bimodal distributions, 204, 205
Biological thought
    in nineteenth century, 2—8
    twentieth-century revolution in, 8—14
    unsolved problems in, 14—28
Bondone, Giotto de, 36
British Broadcasting Company (BBC), 33—34
Brower, Lincoln, 19–21
Brugger, Catalina Aguado, 20
Brugger, Kenneth, 20
Bryant, William Jennings, 170–171
Bueker, Elmer, 120
Butterfly migration, 17–22

C

Cannon, Walter Bradford, 155–158
Carver, George Washington, 165
Case studies
    cholera transmission, 92–101, 187–207
    chromosome counting, 41–44
    discovery of nerve growth factor, 116–122

homing in salmon, 122–138
mass extinctions and end of dinosaurs,
    138–145
Causal hypotheses, 50–57
    cause and effect and, 57
    teleological hypotheses versus, 50–54
    types of causal explanations, 55–57
Cell biology, 11, 12–13
Centers for Disease Control and Prevention
    (CDC), 102, 165–166
Central tendency measures, 201–207
Change of rate, 201
Cholera transmission, 92–101, 187–207
Chromosomes, counting, 41–44
Class, 141
Clinical research, 151
Clockwork of the heavens, 36
Cloning, 164
Cohen, Stanley, 121–122
Columbine High School shootings, 178
Combination drug therapy, 110–111
Common sense, in science, 37
Conceptualization, 39–44
    case study in, 41–44
    types of, 41–42
Conscious bias, 58–59
Control elements, 60, 89–90, 92, 133,
    136, 166
Copernicus, Nicholas, 169, 179
Correlation, 196–200
Correlation coefficient, 199–200
Cosmic dust, 28
Coyne, Jerry A., 177
Creationism, 169–180
    evolution versus, 171–175
    history of, 169–170
    nature of, 170–171
    as science, 175–180
Creativity, in science, 44–45
Crichton, Michael, 73
Crick, Francis, 8, 25
Curtis, Edward, 74–75
Cystic fibrosis, 10

D
Dalai Lama, 180
Darrow, Clarence, 170–171

Darwin, Charles, 3, 8–9, 36–37, 40–41, 45,
    64–65, 67–69, 72, 73, 74, 152–155, 156,
    170, 179
Data analysis and interpretation, 187–207
    central tendency and dispersion, 201–207
    collecting and organizing data, 188–190
    correlation in, 196–200
    presentation of data, 190–196
    rate and change of rate, 201
    sampling error and, 188
Davenport, Charles B., 159
Deduction, 45–47, 48–49
Defoliants, 162–163
Developmental biology, 11
Developmental mechanics, 75–76
Dinosaurs, 138–145
Distribution maps, 190
DNA (deoxyribonucleic acid), 8–10, 69–72, 73
    DNA provirus hypothesis, 70, 71–72
    retroviruses and, 107–109
"Dolly" (cloned sheep), 164
Driesch, Hans, 75–76
Drosophila (fruit fly), 10, 115, 160–162, 174
Drug traffic, 109
Duesberg, Peter, 104

E
Ecocide, 162–163
Ecology, 11, 13–14
Ectoderm, 22–24
Effluvia hypothesis, 93–95, 200
Egerton, Samuel Y., 36
Ehrlich, Paul R., 178, 179
Einstein, Albert, 37, 158
Empirical knowledge, in science, 34–35
Empirical testing, 35, 38–39
Epidermal growth factor (EGF), 121
Essay on Population (Malthus), 44–45, 153–154
Ethical issues, 91, 99
    fraud in research, 57–59
    human medical experimentation, 149–150,
        158–162, 164–168
Eugenics, 159–162
Evolution, 2–3, 8–10, 22–28, 35, 40–41,
    44–45, 64–65, 67–69
    Creationism versus, 171–175
    hens' teeth and, 22–24

origins of life and, 24–28
political economy and, 152–155
Experimental elements, 60, 89–90, 92, 133,
    136, 164–168
Experimentation, 35
    control elements in, 60, 89–90, 92, 133,
        136, 166
    dissection of experiments and, 49–50
    experimental elements in, 60, 89–90, 92,
        133, 136, 164–168
    testing hypotheses by, 88–90, 116–122
External hypothesis, 55–56
Extinction, 138–145
Extrapolation, 194–196

F
Fact
    case study in, 41–44
    in science, 37–39
Family pedigree studies, 159–162
Farley, John, 63–64
Fisher, C., 22–24
Fogel, S., 51–54
Fossilization, 171–174
Fox, Sidney, 26
Fraud, in science, 57–59
French Academy of Sciences, 62–64

G
Gabriel, M. L., 51–54
Galápagos Rift, 14–17
Galileo, Galilei, 169, 179
Galston, Arthur W., 163
Geison, Gerald, 63–64
Generality, in science, 35–36
Generalizations, 40
"Genesis I" hypothesis, 175–177
Genetic engineering, 21–22
Genetics
    eugenics and, 159–162
    Mendelian, 8–10, 150, 159
Genocide, 166–167
Geomagnetic data, 174
Germ theory of disease, 61–66, 92
Goldstein, Inge F., 192, 193
Goldstein, Martin F., 192, 193
Gould, James, 177–178

Gould, Stephen Jay, 179
Graphs, 191–194

H
Haeckel, Ernst, 5
Haldane, J. B. S., 25
Hamburger, Viktor, 76, 77, 119–121
Hansen-Melander, Eva, 43
Hardin, Garrett, 81
Harrel, Ross, 132
Harrison, Ross G., 116–119
Hasler, A. D., 126–128, 132–138
Helmholtz, Hermann von, 2
Helper T cells, 103–105, 106–107
Hens' teeth, 22–24
Herbicides in Southeast Asia, 162–163
Hippocrates, 92
Hippocratic oath, 167
Historical hypothesis, 56–57
HIV (human immunodeficiency virus),
    103–111
Holistic materialism, 78–79
Homeostasis, 155–158
Hooper, Edward, 105
Hsu, T. C., 41
Huff, Darrell, 195
Human Genome Project (HGP), 10
Humanities, science and, 36–37
Human medical experimentation, 149–150,
    158–162, 164–168
Huntington's chorea, 10
Huxley, Thomas Henry, 170
Hypothalamus, 157
Hypotheses, 40
    acceptance of, 48–49
    in dissection of experiments, 49–50
    as explanations, 50–57
    formulation of, 45–46
Hypothesis testing, 47–48, 87–114
    AIDS crisis and, 101–111
    case study in cholera transmission, 92–101
    by experiments, 88–90, 116–122
    by observation, 87–88
    sample size in, 90–92
    uniformity in, 90–92

I

Idealism, 76–79, 181–182
Immunobiology, 11
Impact hypothesis, 139–140, 142
Induction, 45–47, 48–49
Internal hypothesis, 55
Interpolation, 194–196
Intuition, in science, 37
Iridium, 138–140

J

Johanson, Donald, 9
Johnson, George, 13, 108
Juke, Ada, 161
Jurassic Park (film/book), 35, 73

K

Kammerer, Paul, 58–59
Kansas Board of Education, 169, 177, 179–180
Karyotypes, 43–44
Keeton, William, 177–178
Koch, Robert, 92
Kollar, E. J., 22–24
Kuhn, Thomas, 66–67, 72, 73–74, 169

L

Laissez-faire economics, 156–158
Laplace, Pierre Simon de, 181
Latent period, 108–109
Leakey, Louis, 9
Leakey, Mary, 9
Levels of significance, 204–207
Levi-Montalcini, Rita, 119–122
Lewis, G. N., 181
Lillie, F. R., 119
Lincoln, Abraham, 74–75
Line graphs, 192–194
Linnaeus (Karl von Linné), 67, 68–69, 72, 74, 77
Lipid envelope, 106
Loch Ness monster, 35
Logarithmic scales, 194
Logic of science, 37–57
    conceptualization in, 39–44
    deduction in, 45–47, 48–49
    fact and, 37–39
    hypotheses as explanations in, 50–57

hypothesis testing and, 47–48, 87–114
induction in, 45–47, 48–49
observation and, 37–39
predictions in, 47–48
proof and, 48–49
Luyendyk, Bruce, 16
Lysogenic phase, 108–109

M

Macdonald, Ken, 16
Macroevolution, 169
Maggot experiments, 59–61
Malthus, Thomas Robert, 36–37, 44–45, 153–154
Marine squid (*Loligo peali*), 11
Mars, 66, 151
Marx, Karl, 154
Mass extinctions, 138–145
Materialism, 76–79, 181–182
Mayr, Ernst, 54–57
Mean, 201–202
Mechanism, 74–76, 78–79
Median, 202
Mendel, Gregor, 8–10, 150, 159
Mesoderm, 22–24
Microevolution, 170
Microscopes, 42, 80
Midwife toad (*Alytes obstetricans*), 58–59
Miller, Stanley, 25–27
Mode, 202
Molecular biology
    DNA and, 8–10, 69–72, 73, 107–109
    reverse transcriptase and, 69–72, 106–108, 110
    RNA and, 11, 69–72, 73, 107–109
Morgan, Thomas Hunt, 160–162
Morpholine, 132–134
Morphology, 3–6
Mosaic theory, 6
Muller, Hermann J., 162
Müller, Johannes, 80
Muller, Rich, 144

N

Napoleon I, 64, 181
Napoleon III, 64

National Aeronautics and Space
    Administration (NASA), 27–28, 66
National Institutes of Health (NIH), 167
Natural selection, 3, 8–9, 40–41, 44–45,
    64–65, 67–69, 152–155, 176, 180
Nazi Germany, 149–150, 158–162, 164,
    166–167
Neanderthals, 10
Negative factor, 111
Neild, Ted, 33–34
Nemesis hypothesis, 144–145
Nerve growth factor (NGF), 116–122
Neurobiology, 11–12
Neuroblasts, 116–122
Neurons, 116–122
Neurotransmitters, 11–12
New Deal, 158
Newton, Isaac, 36, 182
Nobel, G. K., 59
Null hypothesis, 90, 206

O
Observation
    case study in, 41–44
    in science, 37–39
    testing hypotheses by, 87–88
Olfactory hypothesis for salmon homing,
    126–138
*On the Origin of Species by Means of Natural
    Selection* (Darwin), 3, 24–25, 64–65,
    67–69, 152–155, 170
Oort Cloud, 145
Oparin, A. I., 25
Organelles, 13
Otto, Max, 182–183
Ozone layer, 27

P
Painter, T. S., 42–44
Palm reading, 168
Panspermia hypothesis, 25–28
Paradigms, 66–74
    characteristics of, 72–74
    paradigm shift and, 66–72, 179
Pareto, Vilfredo, 156
Pasteur, Louis, 61–66, 69, 92
Pasteurization, 62

Periodicity, 142–144
Peripatus, 3, 6
Pheromones, 130–134
Phrenology, 168
Phylogeny, 4–5, 6–8
Phylum, 141
Physiology, 2–3
Phytoplankton, 141
Pius IX, Pope, 64
Placebos, 166
Plant growth hormone, 163
Plato, 67–69, 76–77
Polio vaccines, 105–106
Political economy, 36–37, 44–45, 153–155,
    156–158
Pouchet, Felix A., 61–66
Prediction, 47–48
Presley, Elvis, 38
Proof in science, 48–49
Prostitution, 109
Protease inhibitors, 110
Protoplasm, 12–13
Protoplasmic bridge, 116
Provirus, 70, 71
Proximate cause, 181
Pseudoscience, 168–180
    Creationism, 169–180
    examples of, 168
Punctuated equilibrium, 174–175

Q
Qualitative data, 188–189
Quantitative data, 188–189

R
Radioactive dating methods, 174
Rate, 201
Rationality, of science, 35
Raup, David, 142–144
Raven, Peter, 13, 108
Redi, Francesco, 59–61
Relativity, 37
Religion
    Creationism and, 169–180
    difference between science and, 181–182
Repeatability, in science, 35
Retroviruses, 107–109

Reverse transcriptase, 69–72, 106–108, 110
Ribosomes, 13
Ricardo, David, 36–37, 153–154
River, The (Hooper), 105
Rivers, Eunice, 166
RNA (ribonucleic acid), 11, 69–72, 73,
    107–109
Roosevelt, Franklin Delano, 158
Roundworm, 12
Rous sarcoma virus (RSV), 69–71
Roux, Wilhelm, 75–76
Rutherford, Ernest, 80–81

S

Salmon, homing in, 122–138
Sample size, 90–92
Sampling error, 188
Sanford, Scott, 27–28
Scales, 194
Scholz, Allen, 132
Science
    bias in, 57–66
    characteristics of, 34–36
    common sense and, 37
    creativity in, 44–45
    defining, 34
    idealism and materialism in, 76–79, 181–182
    intuition and, 37
    logic of, 37–57
    mechanism and vitalism in, 74–76, 78–79
    paradigms and, 66–74
    relationship between social sciences,
        humanities, and, 36–37
    strengths and limitations of, 80–81
Scopes, John, 170–171
Scopes monkey trial, 170–171
Scott, Eugenie, 172
Sea slug (*Aplysia*), 12
Selvig, Linda, 179–180
Sepkoski, Jack, 142–144
Shorey, M. I., 119, 120
Shumatoff, Alex, 18
SIV (simian immunodeficiency virus), 105
Skehan, James W., 180
Skewed distributions, 204
Smith, Adam, 36–37, 153–154
Snow, John, 92–101, 187–207

Social constructionist view, 151–158
Social context of science, 149–186
    human research subjects and, 149–150,
        158–162, 164–168
    pseudoscience and, 168–180
    religion and, 181–182
    social construction of science, 151–158
    social responsibility of science, 158–164
    technology versus science, 150–151
Social responsibility of science, 158–164
Social sciences, science and, 36–37
Solomon, D. J., 129–131, 134
Spallazani, Lazaro, 50, 51–54, 91
Species, 67–69, 78, 141–142
Spemann, Hans, 6–8
Spontaneous generation hypothesis, 25
    germ theory of disease and, 61–66
    maggots and, 59–61
Standard deviation, 202–204
Statistical significance, 135–138
Stein, Gertrude, 29
Sterilization, 159–162
Structure of Scientific Revolutions, The (Kuhn),
    66–67
Syphilis, 164–168

T

Tables, 191
Technology, science versus, 150–151
Teleological hypotheses, 50–54
Telescopes, 80
Temin, Howard, 69–71, 72–73
Testability, in science, 35, 38–39
Theories, 40–41
Thermal energy, 14–17
Toklas, Alice B., 29
Transmutation of species, 68
Treasure hunt model, 152
Truth tables, 47–48
Tuskegee Study, 164–168
Typhoid fever, 100

U

UFOs, 39
Ultimate cause, 181
Unconscious bias, 59–66
Uniformity, 90–92

United States Public Health Service (USPHS), 164–168
Urey, Harold, 25
Urquart, Fred A., 19–21
Urquart, Nora, 20
Ussher, James, 170

V
Vapor hypothesis of fertilization, 50, 51–54
Variance, 202–204
Vatican I, 64
Vietnam war, 162–163
Vitalism, 74–76, 78

W
Wallace, Alfred Russel, 40–41, 45, 154
Warbler migration, 54–57
Water-borne hypothesis, 95–100

Watson, James D., 8
Whewell, William, 34
Wilberforce, Samuel, 170
Wilson, Edmund Beecher, 12–13
Winniwarter, H. von, 42

X
X chromosomes, 42–43

Y
Young, Robert, 36–37
"Young earth" theory, 174

Z
Zidovudine (AZT), 110
Zooplankton, 141

About the Authors

Garland E. Allen, Ph.D. (Harvard University) is professor of biology at Washington University in St. Louis where he has taught introductory biology to majors and non-majors for over 25 years. He has also taught at Harvard University and has been the Sarton Lecturer at the AAAS. He is the author of *T.H. Morgan, the Man and His Science* (Princeton University Press, 1978), *Life Science in the Twentieth Century* (Cambridge University Press, 1975; 1978), and co-author with Jeffrey J. W. Baker of *The Study of Biology* (Addison-Wesley), a series of basic biology texts for college and university students. He is presently serving as co-editor of the *Journal of the History of Biology*. Dr. Allen is an expert on the history of genetics and the misuse of genetics that resulted in the eugenics movement in the early twentieth century. He lectures to scientific and general audiences and has been involved in teaching courses at the Marine Biological Laboratory in Woods Hole, Massachusetts, where he served as a trustee for eight years, and at the Jackson Laboratory in Bar Harbor, Maine.

Jeffrey J. W. Baker taught for many years at Wesleyan University in Middletown, Connecticut and has held faculty positions at Washington University in St. Louis, The George Washington University, the University of Puerto Rico, and the University of Virginia. Professor Baker is the author or co-author with Dr. Allen of more than a dozen books in the biological sciences, a book on the American Civil War, and two science-related children's books, one of which was selected by the New York Times Book Review as one of the ten best of the year. He has written extensively concerning modern day conflicts between science and religion and was invited to present his views at the Pontifical Academy of Science in the Vatican. Now retired from teaching but still actively writing and lecturing, Professor Baker divides his time between residences in central Virginia and the Island of Vieques, Puerto Rico.